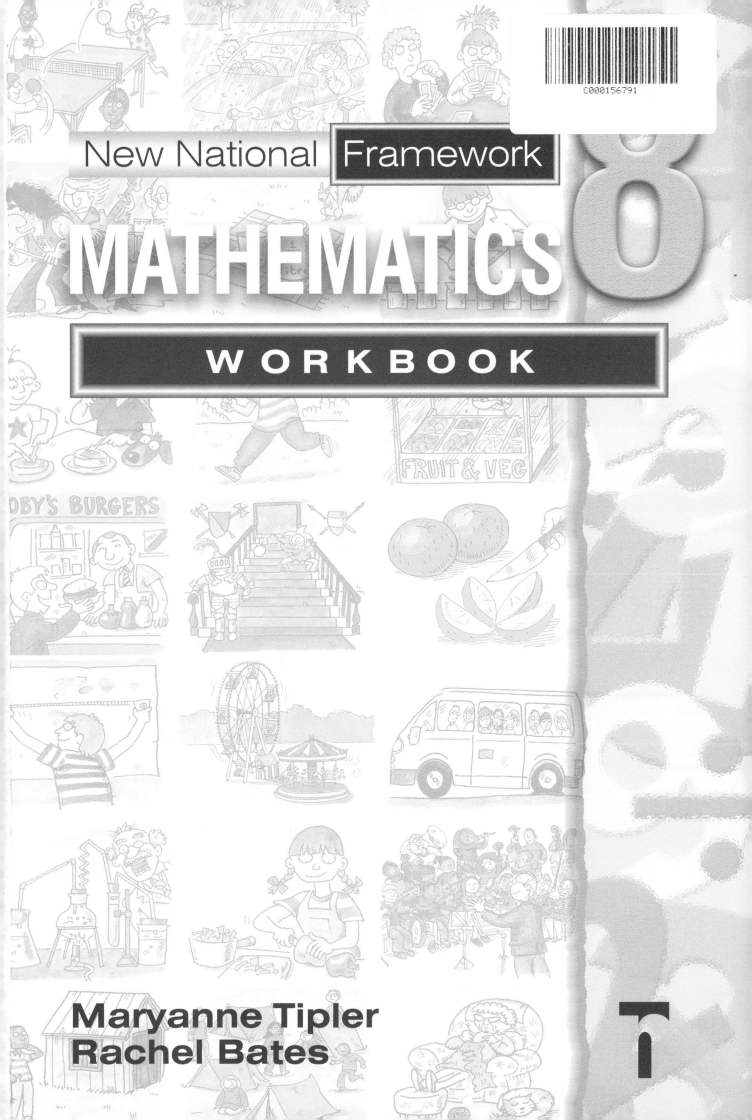

New National Framework

MATHEMATICS 8

WORKBOOK

Maryanne Tipler
Rachel Bates

Published in 2004 by:
Nelson Thornes Ltd
Delta Place
27 Bath Road
CHELTENHAM
GL53 7TH
United Kingdom

04 05 06 07 08 / 10 9 8 7 6 5 4 3 2 1

A catalogue record for this book is available from the British Library

ISBN 0 7487 9137 X

Illustrations by Ian West
Page make-up by Mathematical Composition Setters Ltd

Printed and bound in Croatia by Zrinski

Contents

Contents

Introduction

This workbook has been designed to be used alongside New National Framework Mathematics Core Book 8. However, it can also be used in conjunction with whatever resources are being used or as a fully stand alone resource. It is the ideal course companion to be used at home or at school. All topics from the Framework are covered in an accessible and stimulating format.

The book is divided into four sections. The book contains 105 'topic pages' to match the 105 hours outlined in the Medium-term Plans. Each page has been very carefully developed. At the top of each page are the core objectives to be covered. These are followed by a worked example and/or a summary of the key points. A range of differentiated questions follow, ranging from practice questions to games and puzzles.

Workbooks are also available for the Star and Plus books for all years.

NUMBER

1 Working with Place Value

Let's look at ...
- powers of ten
- writing large numbers in words
- adding and subtracting multiples of 0·1, 0·01 and 0·001

✓ This chart shows the **powers of ten**.

Millions				Thousands			Hundreds		
(Thousands of millions) Billions	Hundreds of millions	Tens of millions	Millions	Hundreds of thousands	Tens of thousands	Thousands	Hundreds	Tens	Units
10^9	10^8	10^7	10^6	10^5	10^4	10^3	10^2	10^1	1

These are the points you need to know.

✓ $6·3 \times 10^4$ is **read as** 'six point three times ten to the power of four'.

✓ **To add or subtract 0·001**, add or subtract 1 to or from the thousandths digit.

    Examples $5·283 - 0·001 = 5·282$ (subtract 1 from the thousandths)
                         $15·999 + 0·001 = 16·000$ (add 1 to the thousandths)

(A) Power on and off

Write these as powers of ten.

a a hundred _____

b ten thousand _____

c ten million _____

Write these without the index.

This little number is called the index.

d 10^5 _____

e 10^7 _____

*f 10^{10} _____

(B) Earth facts

Write the numbers in the boxes in words.

a The surface area of the Earth is [510 000 000] square km. _____

b The area of the Antarctic Ocean is [32 248 000] square km. _____
_____

Write down how you would say the number in the box.

c The temperature of the centre of the Earth is [$4·5 \times 10^3$] degrees centigrade. _____
_____

d The average distance of the Earth from the Sun is [$1·5 \times 10^8$] km. _____
_____

(C) Figure it

Write in figures the number that is

a 2 more than six and a quarter thousand _____

b 3 less than one and a half million. _____

(D) Michael's mistakes

Michael measured the length of some objects in his home and made a mistake with each measurement. Write the true measurements on the table below.

	Object	Michael's Measurement	Mistake	True Measurement
a	Table	2·43 m	0·01 m too short	
b	Bookcase	1·534 m	0·001 m too long	
c	Window	3·158 m	0·006 m too short	
*d	Bath	1·96 m	8 cm too long	

How did you find this? [EASY] [OK] [HARD]

2 Using Place Value to Multiply and Divide

Let's look at ...
● multiplying and dividing by multiples of 10, 100 and 1000
● multiplying and dividing by 0·1 and 0·01

✓ We use **place value** to multiply and divide by multiples of 10, 100, 1000, ...

Examples $1·3 \times 400 = 1·3 \times 4 \times 100$ $3·6 \div 90 = 3·6 \div 9 \div 10$
$= 5·2 \times 100$ $= 0·4 \div 10$
$= 520$ $= 0·04$

These are the points you need to know.

✓ **Multiplying by 0·1** is the same as dividing by 10.

Example $3·6 \times 0·1 = 3·6 \div 10$
$= 0·36$

✓ **Dividing by 0·1** is the same as multiplying by 10.

✓ **Multiplying by 0·01** is the same as dividing by 100.

✓ **Dividing by 0·01** is the same as multiplying by 100.

Example $0·52 \div 0·01 = 0·52 \times 100$
$= 52$

(A) Quick questions

a $50 \times 70 =$ _____
b $300 \times 4000 =$ _____
c $800 \div 20 =$ _____
d $90\,000 \div 300 =$ _____
e $80 \div 4000 =$ _____
f $49 \div 7000 =$ _____
g $2·4 \times 20 =$ _____
*h $0·27 \div 90 =$ _____

(B) Missing numbers

Use a number from the box to complete each equation.

2·2	20	2
200	2000	0·2

a $30 \times \boxed{} = 600$
b $\boxed{} \div 50 = 40$
c $\boxed{} \div 400 = 0·5$
*d $300 \times \boxed{} = 660$

(C) Party time

a Balloons come in boxes of 30. How many balloons are there in
i 40 boxes? _____
ii 500 boxes? _____
iii 1500 boxes? _____

b Poppy buys 24 kg of sweets for her end-of-year tennis party. There are 80 people at the party.

How many kilograms of sweets does each person get?

_____

(D) Matching

Look in the example box at the top for help.

Draw a line to match each expression on the left with an equivalent expression on the right.

a
$3·7 \times 0·1$ •   • $3·7 \div 100$
$3·7 \div 0·1$ •   • $3·7 \div 10$
$3·7 \times 0·01$ •   • $3·7 \times 10$
$3·7 \div 0·01$ •   • $3·7 \times 100$

b
$0·096 \times 0·01$ •   • $0·096 \times 10$
$0·096 \times 0·1$ •   • $0·096 \div 10$
$0·096 \div 0·1$ •   • $0·096 \div 100$
$0·096 \div 0·01$ •   • $0·096 \times 100$

(E) Table time

Fill in this table.

$6 \times 0·1$ $41 \times 0·1$

	6	41	0·46	3·9	0·8	⁻7·29
× 0·1						
÷ 0·1						
× 0·01						
÷ 0·01						

How did you find this?   EASY   OK   HARD

3 Ordering Decimals and Rounding

Let's look at ...
- putting decimals in order
- rounding to powers of ten
- rounding to decimal places

✓ To put **decimals in order**, compare digits with the same place value.

✓ We can round a number to a **power of ten**.

   Example   4 872 487 to the nearest ten thousand is 4 870 000.
   4 872 487 to the nearest hundred thousand is 4 900 000.

✓ To round to a given number of **decimal places**:
   1 keep the number of digits asked for after the decimal point.
   2 delete any following digits. If the first digit to be deleted is 5 or more, increase the last digit kept by 1.

   Examples   8·6832 = 8·7 to 1 d.p.   53·0535 = 53·05 to 2 d.p.

These are the points you need to know.

A Thirteen step maze

The longest path through this maze goes through exactly 13 numbers.

START	0·69	1·34	1·345	1·324	12·301	12·31	12·3	FINISH
	0·7 → 0·71		1·453	1·354	12·03	12·311	12·32	
	0·65	0·708	1·543	1·6	1·61	12·13	12·09	

- Start at 0·7. Always move to a square which has a **larger number**. Make the longest path you can.
- You may move up ↑, down ↓ or right —→. You may **not** move diagonally.

B Quick questions

Round

a   827 to the nearest 10 _____
b   8361 to the nearest 100 _____
c   7835 to the nearest 1000 _____
d   5865 to the nearest 10 _____
e   53 842 to the nearest 10 000 _____
f   317 803 to the nearest 100 000 _____
g   479 500 to the nearest 1000 _____
h   1 845 607 to the nearest million _____

C New York

There are 16 400 parking metres in Manhattan, to the nearest 100.

What is the smallest number of parking metres there could be? _____ largest number? _____

D Corrections

Melanie rounded each of the numbers at the left of this table to 0 d.p., 1 d.p. and 2 d.p. Cross out her mistakes and correct them.

	0 d.p.	1 d.p.	2 d.p.
4·15837	4	4·1	4·16
28·369	29	28·3	28·37
0·9042	0	0·9	0·91
169·598	171	169·5	169·60

E Henrietta's week

Round the answers to these sensibly. Say what you have rounded them to.

a   Henrietta spent 368 minutes on a train over 7 days.
   Calculate her average time spent on a train each day. _____

b   Henrietta's great aunt died. She left £32 250 in her will to be shared evenly between 17 people.
   How much did each person receive? _____

How did you find this?   EASY   OK   HARD

4 Adding and Subtracting Integers

 except  E

Let's look at ...
● adding and subtracting integers mentally
● using a calculator to add and subtract integers

$1 + {}^-1 = 0$ *Example* $45 + {}^-36 = 9 + 36 + {}^-36 = 9 + 0 = 9$

$0 - 1 = {}^-1$ *Example* ${}^-4 - 7 = {}^-4 + 0 - 7 = {}^-4 + {}^-7 = {}^-11$

$0 - {}^-1 = 1$ *Example* $82 - {}^-85 = 82 + 0 - {}^-85 = 82 + 85 = 167$

These are the points you need to know.

A *Did you know?*

	N									N				N		N	
⁻7	**3**		⁻7	0	7	⁻4	6		⁻10	⁻7	**3**	⁻2	⁻3	**3**	7	**3**	⁻2
⁻2	13	7	⁻4	7		⁻3	5		⁻26		⁻10	⁻3	⁻2	6			
⁻10	⁻26	10	10	7	⁻20		⁻4	⁻7	⁻15	7							

Write the letter beside each question above its answer in the box.

N $4 + {}^-1 = 3$ **V** $5 + {}^-5$ **T** $6 + {}^-8$ **S** ${}^-4 + 9$

I ${}^-1 + {}^-2$ **H** $9 - {}^-4$ **Y** ${}^-2 - {}^-8$ **A** ${}^-15 + {}^-11$

O $18 + {}^-25$ **R** ${}^-36 - {}^-32$ **E** $5 - 6 + 8$ **C** $50 + {}^-20 + {}^-40$

L ${}^-30 + 50 + {}^-40 + 30$ **D** ${}^-40 - 20 + 15 + 25$ **M** $18 - 36 - 9 + 12$

B *Pyramids*

a

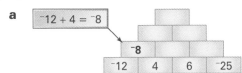

${}^-12 + 4 = {}^-8$

⁻8

⁻12 4 6 ⁻25

Add each pair of numbers to get the number above.

b

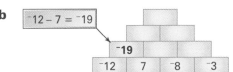

${}^-12 - 7 = {}^-19$

⁻19

⁻12 7 ⁻8 ⁻3

Subtract the number on the right from the number on the left to get the number above.

C *Missing numbers*

Complete each of these.

a ${}^-6 + \boxed{} = 2$ **b** ${}^-3 - \boxed{} = {}^-8$ **c** ${}^-5 - \boxed{} = 2$ **d** $19 + \boxed{} = {}^-7$

＊Complete the following in three different ways.

e $\boxed{} - \boxed{} = {}^-12$ $\boxed{} - \boxed{} = {}^-12$ $\boxed{} - \boxed{} = {}^-12$

＊D *Star puzzle*

Place each of the numbers below in one of the circles. Each line in the star must **add up to zero**.

${}^-5, {}^-4, {}^-3, {}^-2, {}^-1, 0, 1, 2, 3$

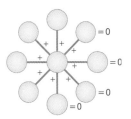

E *Calculator time*

Use a calculator to find these.

a $371 + {}^-193 = \underline{}$ **b** ${}^-123 + {}^-463 = \underline{}$ **c** $179 - {}^-148 = \underline{}$

d $17{\cdot}8 - {}^-42{\cdot}6 = \underline{}$ **e** ${}^-58 - 104 - {}^-79 = \underline{}$ **f** ${}^-26 + {}^-269 - {}^-108 = \underline{}$

How did you find this? EASY OK HARD

5 Multiplying and Dividing Integers

  except **E**

Let's look at ...
● multiplying and dividing integers mentally
● using a calculator to multiply and divide integers

These are the points you need to know.

✓ **Multiplying (or dividing) two negative numbers** gives a positive number.

✓ **Multiplying (or dividing) one negative and one positive number** gives a negative number.

Examples $^-5 \times {}^-7 = 35$ $^-4 \times 6 = {}^-24$ $^-24 \div 8 = {}^-3$ $^-32 \div {}^-4 = 8$

A Odd one out

Find each of the following and circle the odd one out in each line.

a $5 \times {}^-3 =$ ____ $^-3 \times {}^-5 =$ ____ $30 \div {}^-2 =$ ____ $1 \times {}^-15 =$ ____

b $^-4 \times {}^-6 =$ ____ $3 \times 8 =$ ____ $^-48 \div {}^-2 =$ ____ $^-2 \times 12 =$ ____

c $^-9 \times {}^-1 =$ ____ $^-27 \div 3 =$ ____ $3 \times {}^-3 =$ ____ $81 \div {}^-9 =$ ____

B Number chains

Complete these number chains. $24 \div {}^-3 = {}^-8$

a 24 $\div {}^-3$ ⁻8 $\div {}^-4$ ☐ $\div {}^-2$ ☐ $\times {}^-6$ ☐

b ⁻6 $\times 5$ ☐ $\div {}^-2$ ☐ $\div {}^-3$ ☐ $\times {}^-2$ ☐

*c ☐ $\times {}^-4$ ☐ $\div 5$ 8 $\div {}^-2$ ☐ $\div {}^-4$ ☐

C Grid time

Complete these multiplication grids.

a

×	3	⁻1	⁻4
⁻2	⁻6	2	
5			
⁻7			

b

×	5	⁻2	⁻9
⁻6			
⁻4			
3			

c

×			⁻9
⁻5	⁻15	35	
			⁻28
	⁻24		

D Four questions

The answer is ⁻12.
Write down two multiplications and two divisions with this answer.

_____   _____   _____   _____

E Cross number

Use your calculator to complete this cross number.

Across
4. $12 \cdot 2 \times {}^-16 = {}^-\mathbf{195 \cdot 2}$
5. $^-1860 \div {}^-124$
6. $^-734 \cdot 88 \div {}^-24$
7. $^-722 \cdot 4 \div 8 \cdot 6$
9. $1225 \div {}^-14$

Down
1. $^-646 \cdot 8 \div 7$
2. $^-0 \cdot 25 \times {}^-0 \cdot 08$
3. $^-26 \times 2 \cdot 9$
4. $948 \div {}^-4$
7. $^-12 \cdot 5 \times 4 \cdot 4$
8. $^-562 \cdot 8 \div {}^-8 \cdot 4$

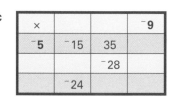

How did you find this?   EASY   OK   HARD

6 Order of Operations with Integers

Let's look at ...
● solving problems with integers and more than one operation

These are the points you need to know.

✓ The **order** in which we do **operations** is

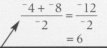
Remember **BIDMAS** and this will help you.

Brackets → **I**ndices → **D**ivision and **M**ultiplication → **A**ddition and **S**ubtraction

Examples $^-7 - 2 \times {}^-3 = {}^-7 - {}^-6$
$= {}^-1$

Multiply first
$2 \times {}^-3 = {}^-6$

$^-4({}^-6 + 1) = {}^-4 \times {}^-5$
$= 20$

Brackets first
$^-6 + 1 = {}^-5$

$\dfrac{{}^-4 + {}^-8}{{}^-2} = \dfrac{{}^-12}{{}^-2}$
$= 6$

Remember the division line acts as a bracket.

A Iceberg ahead

Find the answers to each of the following, then shade them in the grid.
The **largest two numbers** left at the end are the length and width (in km) of the largest recorded iceberg.

a $^-4 + 3 \times 6 = \mathbf{14}$

b $2 \times {}^-5 + 3 = \_\_\_\_$

c $^-10 - {}^-4 \times {}^-1 = \_\_\_\_$

d $1 + {}^-200 \times {}^-2 = \_\_\_\_$

e $4(6 + {}^-7) = \_\_\_\_$

f $^-3({}^-20 - 5) = \_\_\_\_$

g $100 + {}^-2 \times {}^-4 = \_\_\_\_$

h $200 - 4 \times {}^-7 = \_\_\_\_$

i $5 \times {}^-2 + 6 \times {}^-1 = \_\_\_\_$

j $^-9 \times {}^-6 + {}^-3 \times 4 = \_\_\_\_$

k $4({}^-8 - {}^-2) + 5 = \_\_\_\_$

l $\dfrac{8 + {}^-2}{{}^-3} = \_\_\_\_$

m $\dfrac{3({}^-10 + 2)}{2} = \_\_\_\_$

n $\dfrac{{}^-6(3 - 8)}{{}^-10} = \_\_\_\_$

401	92	3	12	$^-3$
$^-75$	108	$^-52$	$^-16$	$^-19$
$^-30$	2	335	$^-4$	$^-14$
14	42	$^-7$	$^-2$	36
228	75	72	97	45
$^-94$	$^-6$	32	6	$^-12$

The largest recorded iceberg was _____ km by _____ km.

B Calculator confusion

Andy and Judy each worked out some calculations on their calculators and got different answers. Decide who has the correct answer each time.

	Calculation	Andy's Answer	Judy's Answer	Who is correct?
a	$4 + 2 \times {}^-3$	$^-2$	$^-18$	
b	$^-5 + {}^-7 \times 2$	$^-19$	$^-24$	
c	$^-20 - 5 \times {}^-3$	$^-5$	75	
d	$18 - 12 \div {}^-6$	20	$^-1$	

e Who made the most errors? _____   Explain what he/she was doing wrong. _____

_____

* C Missing operations

Use +, −, × or ÷ signs to make these true.

a $4 \boxed{\times} {}^-3 \ \Box \ 2 = {}^-10$

b $^-3 \ \Box \ {}^-5 \ \Box \ 1 = 14$

c $12 \ \Box \ 6 \ \Box \ {}^-3 = 30$

d $^-30 \ \Box \ {}^-12 \ \Box \ 2 = {}^-24$

Place one set of brackets in each statement to make these true.

e $^-6 + {}^-4 \times 2 = {}^-20$

f $9 - 8 \times {}^-11 = {}^-11$

g $^-7 + 5 \times {}^-8 - 2 = {}^-57$

7 Divisibility, Factors and Multiples

Let's look at ...
- divisibility by numbers larger than 10
- writing numbers as a product of prime factors
- finding the HCF (highest common factor) and LCM (lowest common multiple) of two numbers

These are the points you need to know.

✓ **Divisibility**

  Example **To check if a number is divisible by 24**, check for divisibility by 3 and 8, not 4 and 6. We check by dividing by two numbers which multiply to 24 but which have no common factor other than 1.

✓ A **prime factor** is a factor that is a prime number. We can use a table or factor tree to write 180 as a product of primes.
  $$180 = 2 \times 2 \times 3 \times 3 \times 5$$
  $$= 2^2 \times 3^2 \times 5$$

Divide by the smallest prime possible each time.

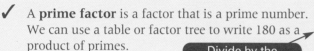

There are other ways to draw a factor tree for 180.

✓ We can use a diagram to find the HCF (highest common factor) or LCM (lowest common multiple) of two numbers.

  Example $360 = 2^3 \times 3^2 \times 5$ $288 = 2^5 \times 3^2$
  HCF of 360 and 288 $= 2 \times 2 \times 2 \times 3 \times 3 = 72$
  LCM of 360 and 288 $= 5 \times 2 \times 2 \times 2 \times 3 \times 3 \times 2 \times 2 = 2^5 \times 3^2 \times 5 = 1440$

A *A or B?*

Which numbers would you use to check divisibility by 12? **A** 3 and 4   **B** 2 and 6 _____
Explain. _____

B *Pick of the bunch*

Which of the numbers in the box are divisible by

1440	2040	13 940
19 836	9135	9240

a 12 _____   **b** 15 _____
c 18 _____   **d** 20 _____

C *Break it down*

Fill in the missing numbers.

a

2	150
	75
	25

b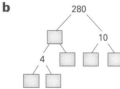

c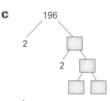

Write each of these as a product of prime numbers in index notation.

d 150 = _____   **e** 280 = _____   **f** 196 = _____

D *Fill it in*

Complete each diagram, then find the HCF and LCM.

a

HCF = _____
LCM = _____

b

HCF = _____
LCM = _____

c

HCF = _____
LCM = _____

How did you find this? EASY OK HARD

8 Squares and Cubes

Let's look at ...
- finding squares and square roots of numbers
- finding cubes and cube roots of numbers

✓ When we find a **square root** there is a positive and a negative answer.

    Example $\sqrt[\pm]{25} = 5$ or $^-5$ because $5 \times 5 = 25$ *and* $^-5 \times ^-5 = 25$.

> When we want the positive **and** negative square root given we write $\sqrt[\pm]{\ }$

✓ We can find the square root of some numbers by **factorising**.

    Example $\sqrt{225} = \sqrt{9 \times 25} = \sqrt{9} \times \sqrt{25} = 3 \times 5 = 15$

> These are the points you need to know.

✓ We can estimate square roots by finding upper and lower bounds.

    Example $\sqrt{13}$ $\sqrt{9} < \sqrt{13} < \sqrt{16}$ so $3 < \sqrt{13} < 4$

✓ $1^3 = 1$ $2^3 = 8$ $3^3 = 27$ $4^3 = 64$ $5^3 = 125$ $10^3 = 1000$

    1, 8, 27, 64, 125 and 1000 are all **cube numbers**.

    $\sqrt[3]{1} = 1$ $\sqrt[3]{8} = 2$ $\sqrt[3]{27} = 3$ $\sqrt[3]{64} = 4$ $\sqrt[3]{125} = 5$ $\sqrt[3]{1000} = 10$

    $\sqrt[3]{8}$ is a **cube root**.

(A) Who is right?

Decide who has got each question correct.

	a $(7-2)^2$	b $(^-9)^2$	c $\sqrt{57+7}$	d $\sqrt[\pm]{100-19}$	e $\sqrt{90-55+1}$	f $(18-3-4)^2$	g $\frac{(12-8)^2}{(1-3)^2}$
Annie	49	$^-81$	8	81	7	121	4
Bella	25	99	64	10	6	110	$^-2$
Carl	5	81	50	9 or $^-9$	5	144	2
Who is right?							

(B) Pluses and minuses

a Give the positive and negative square roots of

    i 16 _____   **ii** 64 _____   **iii** 36 _____

b Find   **i** $\sqrt[\pm]{49}$ **ii** $\sqrt[\pm]{100}$

(C) Missing numbers

Use factorising to fill in these missing numbers.

a $\sqrt{225} = \sqrt{\boxed{} \times 25} = \boxed{} \times 5 = \boxed{}$ ***b** $\sqrt{196} = \sqrt{\boxed{} \times \boxed{}} = \boxed{} \times \boxed{} = \boxed{}$

(D) Quick questions

a 2 cubed equals ____.   **b** ____ is the cube of 3.   **c** 1 is the cube of ____.

d The cube root of 64 is ____.   **e** 10 cubed equals ____.   **f** ____ is the cube root of 125.

g $\sqrt[3]{27} =$ ____   **h** $1^3 =$ ____   **i** $\sqrt[3]{1000} =$ ____   **j** $(^-2)^3 =$ ____   ***k** $(0{\cdot}2)^3 =$ ____

(E) Calculator time

> Give your answers to 2 d.p.

Use your calculator to find these squares, square roots, cubes and cube roots.

a $8{\cdot}4^2 =$ _____   **b** $\sqrt{194} =$ _____   **c** $\sqrt{647-159} =$ _____   **d** $\sqrt{17^2 - 13^2} =$ _____

e $\dfrac{1{\cdot}8^2 + 2{\cdot}7^2}{7+2} =$ _____   **f** $(^-16)^3 =$ _____   **g** $2{\cdot}5^3 =$ _____   **h** $\sqrt[3]{232} =$ _____

How did you find this?   [EASY]   [OK]   [HARD]

9 Adding and Subtracting Mentally

Let's look at ...
● using a variety of mental strategies to add and subtract

✓ These strategies can be used to **add and subtract mentally**:
  ● Complements in 1, 10, 50, 100 and 1000
  ● Partitioning
  ● Counting up
  ● Nearly numbers
  ● Adding and subtracting too much then compensating
  ● Using facts you already know

These are the points you need to know.

A Quick questions

Answer all of the questions on this page mentally. You may use jottings.

a 5 + ¯4 + ¯2 + 7 = _____

b 11 + 3 + ¯8 + ¯6 + 7 = _____

c 0·5 + 0·6 + 0·7 = _____

d ¯1·6 − 0·7 + 0·3 + 0·9 = _____

e 7·2 − 0·9 = _____

f 5·2 + 7·8 = _____

g 0·54 + 0·85 = _____

h 0·73 − 0·48 = _____

i 42 + 58 + 28 + 22 = _____

j 0·31 + 3 + 0·69 − 2 = _____

k 324 + 98 = _____

l 4215 − 504 = _____

m 7·53 + 8·4 = _____

n 5·03 − 2·7 = _____

B Odd one out

A

¯12 21
4
8 ¯13

B

0·3 1·6
1·5
¯0·9 0·5

C

52 31
150
19 48

D

0·42 ¯5
2·02
7 ¯0·3

Which diagram is the odd one out? _____ Explain why. _____

_____

C Magic squares

Complete these magic squares so that every row, column and diagonal adds to the same number.

a

		21	
29			3
	17	15	
7	13	19	25

b

	4·2	6·7	9·2
8·7		3·7	
		5·2	2·7
3·2		6·2	7·7

c

¯7			3
7	4	¯3	¯6
		1	
¯4			2

*D Puzzle time

What are my two numbers if

a the sum is 34 and the difference is 12 _____ and _____

b the sum is 3·9 and the difference is 0·3 _____ and _____

c the sum is ¯3 and the difference is 11 _____ and _____

d the sum is 307 and the difference is 103? _____ and _____

How did you find this? EASY   OK   HARD

10 Multiplying and Dividing Mentally

Let's look at ...
● using a variety of mental strategies to multiply and divide

 These strategies can be used to **multiply and divide mentally**.
- **Multiplying and dividing by multiples of 10 and 100**
- **Place value**
- **Partitioning**
- **Factors**
- **Near tens**
- **Known facts**
- **Doubling and halving**

These are the points you need to know.

A Getting around

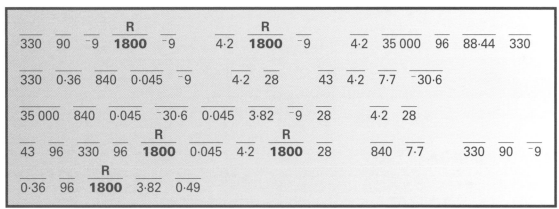

		R			R					
330	90	⁻9	**1800**	⁻9	4·2	**1800**	⁻9	4·2	35 000	96	88·44	330

330 90 ⁻9 **1800** ⁻9   4·2 **1800** ⁻9   4·2 35 000 96 88·44 330

330 0·36 840 0·045 ⁻9   4·2 28   43 4·2 7·7 ⁻30·6

35 000 840 0·045 ⁻30·6 0·045 3·82 ⁻9 28   4·2 28

43 96 330 96 **1800** 0·045 4·2 **1800** 28   840 7·7   330 90 ⁻9

0·36 96 **1800** 3·82 0·49

Write the letter beside each question above its answer in the box. Find the answers mentally.

R 60 × 30 = **1800**   **B** 50 × 700   **H** 2700 ÷ 30   **O** 24 × 4   **I** 168 × 5
T 22 × 15   **A** 7 × 0·6   **W** 0·04 × 9   **C** 0·09 ÷ 2   **S** 5·6 × 5
E ⁻1·8 × 5   **U** 22 × 4·02   **M** 344 ÷ 8   **L** 7·64 ÷ 2   **D** 2·45 ÷ 5
N 2·2 × 3·5   **Y** ⁻6·8 × 4·5

B Wordy wonderings

a Find the product of 24·3 and 0·1. _____

b Add 6 to the product of 3·6 and 3. _____

c Find the quotient when 6·32 is divided by 4. _____

*d Add the quotient of 390 and 6 to the product of 6·5 and 16. _____

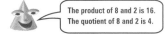 The product of 8 and 2 is 16. The quotient of 8 and 2 is 4.

* C Make it true

For each question, use the numbers in the box to make the calculation true.

a | 5 | 120 | 130 | ☐ × ☐ + ☐ = 770

b | 1·6 | 2·4 | 3·5 | ☐ × ☐ + ☐ = 10

c | 6 | 8 | 152 | ☐ ÷ ☐ − ☐ = 13

d | 0·1 | 0·3 | 0·5 | 4·2 | ☐ × ☐ + ☐ × ☐ = 0·57

 Remember BIDMAS for this question.

11 Solving Problems Mentally

Let's look at ...
● solving word problems mentally

These are the points you need to know.

✓ We can use **mental strategies to solve problems**.
On this page you will need to decide whether to add, subtract, multiply or divide.
In many questions you will need to do more than one of these.

A **Quick questions** Remember the units.

Write the answers to these as quickly as possible.

a How many grams are there in 8·2 kilograms? _____

b How many millimetres are there in 100·2 centimetres? _____

c How many hours are there in 5 days? _____

d Calculate $(15 - 8)^2$. _____

e What is x if $124 + x = 150$? _____

f Two angles in a triangle are 55° and 65°. What is the third angle? _____

g Six pens cost £1·80. What does each one cost? _____

h What is the mean of 8, 15 and 22? _____

B **Perplexing problems**

a Dale has these three cards: 2, 3, 4. He sets them out like this: ☐ ☐ × ☐

He could get 4 2 × 3. ◄— The **answer** to this is **126**.

Find the other five possible answers. _____, _____, _____, _____, _____

b Charlie held his breath for 1·9 minutes.
How many seconds is this? _____

c The Court Theatre seats 400 people in 16 rows.
If each row has the same number of seats, how many seats are in each row? _____

d The Hamilton family consists of Mum, Dad and four children.
They all went to the Kingsgate Fun Park for the day.
If Mrs Hamilton handed over £100 at the entrance booth, how much change did she get? _____

KINGSGATE
· FUN PARK ·
Entrance Fee
Adults   £10.65
Children   £7.80
(some rides cost extra)

e 1 gallon is about 4·5 litres.
How many litres of petrol did each of these people buy?

i 0 6 · 0 gallons

Bill's car = ____ litres

ii 0 0 · 5 gallons

Jim's lawnmower = ____ litres

iii 1 1 · 0 gallons

Nancy's truck = ____ litres

f I have 12 coins in my pocket.
There are some 20p coins, some 50p coins, and nothing else.
Altogether I have £3·60.
How many 20p coins do I have? _____

How did you find this?   EASY   OK   HARD

12 Making Estimates

Let's look at ...
● deciding if an estimate or an exact number is required
● estimating the answer to a calculation by rounding the numbers

These are the points you need to know.

✓ Sometimes **an estimate** is a good enough answer to a question. We make the **best estimate** possible.

Example The number of people who attended a concert?

✓ To **estimate the answer to a calculation** we round the numbers. There is often more than one possible estimate.

Example $382 \times 24 \approx 400 \times 20 = 8000$
or $382 \times 24 \approx 400 \times 25 = 10\ 000$
or $382 \times 24 \approx 380 \times 20 = 7600$

The best estimate is one close to the answer but still easy to do in your head.

✓ Try to **round to 'nice numbers'** when estimating.

Example Approximate $\frac{61}{7}$ to $\frac{63}{7}$ rather than to $\frac{60}{7}$.

63 is a multiple of 7.

A Exact or not?

Decide whether each of the following need an **exact number**, or just an **estimate**.

a A school secretary is organising buses for an outing. She wants to know the number of students who will be attending the outing. _____

b A reporter wants to know how many people attended the Chelsea Flower Show. _____

c A fire officer wants to know how many people were in a burning office block. _____

d A boarding kennel owner wants to know how much food a dog eats each day. _____

B Who's the best?

Christine, Julia and Elizabeth each write down how they would estimate some calculations. Decide whose estimate is the best, for each calculation.

	Calculation	Christine	Julia	Elizabeth	Best Estimator
a	246×92	250×100	250×90	240×90	
b	$581 \div 23$	$600 \div 20$	$600 \div 30$	$550 \div 30$	
c	$19{\cdot}3 \times 39{\cdot}96$	19×39	20×40	20×30	

C Your turn

Estimate the answers to these. Show how you found your estimate.

a $417 \times 29 =$ _____ × _____ = _____   **b** $769 \div 41 =$ _____ ÷ _____ = _____

c $287 \times 64 =$ _____   **d** $14{\cdot}7 \times 4{\cdot}2 =$ _____

e $55{\cdot}2 \div 6{\cdot}8 =$ _____   **f** $305 \div 5{\cdot}7 =$ _____

g A farmer needs 1·8 m of wood for each fence post. About how much wood does he need for 204 posts? _____

h Shelley drinks 23 bottles of juice in a week. Each bottle contains 675 mℓ of juice. About how many millilitres of juice does she drink in a week? _____

i Carl paid £297·35 for 19 computer games, on behalf of his classmates. About how much should he charge each class member for one computer game? _____

13 Adding and Subtracting – Written Calculations

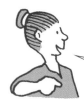

Let's look at ...
- adding and subtracting decimal numbers

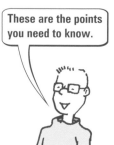

These are the points you need to know.

✓ When we **add and subtract decimals** we line up the decimal points.

Example 53·8 – 6·24 + 3·8 is approximately equal to 54 – 6 + 4 = 52.

```
  4 1 7 1
   53·80          47·56
 –  6·24        +  3·8
   47·56          51·36
                  1 1
```

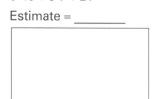

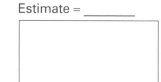

Always estimate the answer first.

(A) *Show your working*

Estimate and calculate each answer. Show your working in the box.

a 3·81 + 42·8 + 5·4
Estimate = ________

b 0·43 + 8·7 + 21
Estimate = ________

c 17 + 0·85 + 1·8 + 11·64
Estimate = ________

d 15 – 3·41 – 2·7
Estimate = ________

e 46·7 – 20·35 + 6·8 – 0·37
Estimate = ________

(B) *Ross's day*

a Ross wanted to know the mass of his dog.
He stood on his digital scales holding the dog. Their combined weight was 101·8 kg.
On his own Ross weighed 68·92 kg.
How heavy was his dog? ____________

b Ross made a fruit cocktail from 1·8 ℓ of apple juice, 3·6 ℓ of pineapple juice and 2·25 ℓ of orange juice.
He filled a 5 ℓ jug with the cocktail.
How much extra cocktail did he have? ____________

(C) *Missing digits*

Fill in the missing digits to make these true.

a
```
    □ · 2
  4 · □ 7
+ 5 · 9 □
□ 3 · 2 3
```

b
```
    0 · 4 □
    6 · □
+ 2 □ · 1 9
  □ 5 · 9 8
```

c
```
    7 0 · □ □
–  □ □ · 2 7
    5 3 · 8 9
```

***d** 64·□ + 3·8□ + 1□ = □0·32

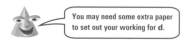

You may need some extra paper to set out your working for d.

How did you find this? EASY OK HARD

14 Multiplying – Written Calculations

Let's look at ...
● multiplying by a decimal number

These are the points you need to know.

✓ **Multiplication**   Always estimate first.

Example   83·2 × 3·4

83·2 × 3·4 is approximately equal to 100 × 3 = 300.

Method 1	**Method 2**
	83·2 × 3·4 is equivalent to 83·2 × 10 × 3·4 × 10 ÷ 100 or 832 × 34 ÷ 100.

Method 1

×	80	3	0.2	check
3	240	9	0·6	249.6
0·4	32	1·2	0·08	+ 33·28
	272 + 10·2 + 0·68			**282·88**

Method 2

83·2 × 3·4 is equivalent to 83·2 × 10 × 3·4 × 10 ÷ 100
or 832 × 34 ÷ 100.

```
    832
×    34
  24960    30 × 832
   3328     4 × 832
  28288
```

Check the answer is about the right size by looking at the estimate.

Answer   28 288 ÷ 100 = **282·88**

(A) Show your working

Estimate and calculate each answer. Show your working in the box.

a 362 × 0·7

Estimate = _____

b 9·3 × 235

Estimate = _____

c 3·84 × 5·2

Estimate = _____

(B) Grid work

You will need some extra paper for working for activities B, C and D.

Complete these multiplication grids. Estimate first.

a

×	36	0·8	2·3
29			
607			
27·5			

b

×	0·6	3·5	8·7
2·6			
71·3			
6·54			

(C) Beth's barbecue

Beth bought
● 4·5 kg of sausages at £2·35 per kg
● 3·2 kg of chops at £6·95 per kg
● 2·7 kg of steak at £8·19 per kg.

Round your answer to the nearest penny.

What was the total cost of the meat for her barbecue? _____

*(D) Puzzle it out

1	2
8	7

Choose from the numbers in the box to work out what values A, B, C and D have.

AB·CD
  × 4
DC·BA

A = _____
B = _____
C = _____
D = _____

15 Dividing – Written Calculations

Let's look at ...
● dividing by a whole number
● dividing by a decimal number

✓ **Division**

Example $89\cdot6 \div 23$

$89\cdot6 \div 23$ is approximately equal to $100 \div 25 = 4$.

Always estimate first.

If we want the answer to 1 d.p. we need to work out the answer to 2 d.p. Then we round to 1 d.p.

```
23)89·6
   69·0        23 × 3
   20·6
   18·4        23 × 0·8
    2·20
    2·07       23 × 0·09
    0·13

Answer  3·89 R 0·13
        3·9 (1 d.p.)
```

These are the points you need to know.

✓ When we **divide by a decimal** we do an equivalent division.

Examples $54 \div 0\cdot9$ is equivalent to $540 \div 9$.
$57\cdot2 \div 3\cdot4$ is equivalent to $572 \div 34$.

$$\frac{54}{0\cdot9}{\,}^{\times 10}_{\times 10} = \frac{540}{9}$$

(A) *Show your working*

Find the answers to these problems. Use the boxes provided to show your working.
Give the answers to **b** and **d** to 1 d.p.

a $429 \div 22$	b $53\cdot7 \div 19$	c $455 \div 0\cdot7$	d $387 \div 2\cdot3$

(B) *Who was Mrs Bandaranaike?*

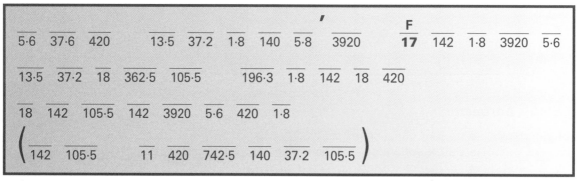

```
                                                                    '                              F
5·6   37·6   420         13·5   37·2   1·8   140   5·8   3920       17   142   1·8   3920   5·6

13·5   37·2   18   362·5   105·5        196·3   1·8   142   18   420

18   142   105·5   142   3920   5·6   420   1·8

( 142   105·5        11   420   742·5   140   37·2   105·5 )
```

Calculate these to 1 d.p. when rounding is necessary.
Write the letter beside each question above its answer in the box.

You will need some extra paper for the working.

F $255 \div 15 = \mathbf{17}$	**M** $396 \div 22$	**C** $308 \div 28$	**W** $216 \div 16$
T $95\cdot2 \div 17$	**O** $521 \div 14$	**H** $677 \div 18$	**D** $93\cdot3 \div 16$
R $58\cdot6 \div 32$	**L** $56 \div 0\cdot4$	**E** $252 \div 0\cdot6$	**S** $196 \div 0\cdot05$
I $355 \div 2\cdot5$	**A** $580 \div 1\cdot6$	**Y** $594 \div 0\cdot8$	**P** $471 \div 2\cdot4$
N $348 \div 3\cdot3$			

How did you find this?   EASY   OK   HARD

16 Checking Answers

Let's look at ...
● using different methods to check the answer to a calculation

These are the points you need to know.

✓ We can **check the answer to a calculation** in one of these ways.
● Check the answer is sensible.
● Check the answer is about the right order of magnitude.
● Estimate first then check the order of magnitude of the answer is the same as the estimate.
● Check using inverse operations, using an equivalent calculation or by checking the last digits.

(A) *Is Jack right?*

Is Jack's comment sensible? Explain why or why not.

a Jack bought 3 milkshakes at £1·95 each and gave the shop assistant a £10 note.
He said, '**I should get £4·15 change.**' _____

_____

b Jack had a bag of 589 sweets. He said, 'If I share them into three equal piles,
there will be 213 sweets in each pile.' _____

_____

c Jack's rugby team scored 43, 26, 17 and 32 points in their last four games.
Jack said, 'The mean number of points is 51.' _____

_____

(B) *True or false?*

a 46 × 0·7 is bigger than 46. _____   **b** 46 × 1·2 is bigger than 46. _____

c 46 ÷ 0·8 is bigger than 46. _____   **d** 250 ÷ 50 is equivalent to 500 ÷ 25. _____

e 250 ÷ 50 is equivalent to 250 ÷ 10 ÷ 5. _____   **f** 250 ÷ 50 is equivalent to 125 ÷ 25. _____

(C) *Check them*

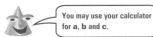 You may use your calculator for a, b and c.

Check these by doing an inverse calculation. Write down the calculation you did.

a 20·24 − 16·49 = 3·75 _____

b 134 × 2·3 = 308.2 _____

c $9·8^2 = 96·04$ _____

Check these by looking at the last digits. Circle any which are wrong.

d 72 × 53 = 3816   **e** 6·4 × 5·6 = 35·82   **f** 0·9 × 127 = 114·8

(D) *Your choice*

Without using a calculator, choose a possible answer to each calculation. Explain your choice.

a 29 × 38   **A** 1462   **B** 1102   **C** 1256
_____ because _____

_____

b 563 × 1·34   **A** 416·82   **B** 609·75   **C** 754·42
_____ because _____

_____

17 Using the Calculator

Let's look at ...
- **using brackets on a calculator**
- **using the calculator memory**

> These are the points you need to know.

✓ When **brackets** are part of a calculation, we key them as we come to them.

  Example $(5·6 + 2·3) × (8·4 − 6·9)$ is keyed as

  (5·6 + 2·3) × (8·4 − 6·9) = to get 11·85.

✓ Sometimes we need to **add brackets to the calculation**.

  Example $\frac{5 + 11}{18 − 7}$ is keyed as

  (5 + 11) ÷ (18 − 7) = to get 1·45. (2 d.p.)

> The whole numerator must be divided by the whole denominator.

✓ We use the **calculator memory** to store an answer.

  STO M+   stores the number on the screen in the memory.

  M+   adds the number on the screen to the number already in the memory.

  RCL M+   recalls the number that is in the memory back to the screen.

> We use 0 STO M+ to clear the memory before beginning a calculation.

(A) Whoops!

a Sarah keyed $\frac{7 × 8}{9 − 6}$ as (7 × 8) ÷ (9 − 6 =

What mistake did Sarah make? _____

b Naim keyed $\frac{3 + 4 − 2}{8 × 6}$ as (3 + 4 − 2) ÷ 8 × 6 =

What mistake did Naim make? _____

(B) Cross number

Complete this crossnumber.

¹1	²3	.	³4		⁴
			⁵		
⁶					
⁷					
			⁸		

Across
1. $5·2 + (3·6 − 1·9) + (8 − 1·5) =$ **13·4**
5. $\frac{145 × 80}{4^2}$
6. $\sqrt{12·96} + 4 × 1·28$
7. $448·5 ÷ \{(9 − 6) × 5\}$
8. $17 × \{18 − (17 − 3)\}$

Down
2. $(20·7 − 3·8) × (12·6 − 10·5)$
3. $503·2 − 6·3(5·7 − 0·9)$
4. $\frac{49 + 76}{8}$ to 2 d.p.

(C) Store, add and recall

> Always use 0 STO M+ to clear the memory before beginning a calculation.

Use the memory keys on your calculator to answer these.

a Ashleigh bought four T-shirts at £12.37 each and three pairs of shorts at £16.54 each.
How much change did she get from £100? _____
Write down the rest of the keying sequence you used.

0 STO M+ _____

b A marathon is 42·2 km long. Matthew and his friends decided to run a marathon as a team.
Three girls each ran 5·7 km and four boys each ran 4·9 km.
What distance was left for Matthew to run? _____
Write down the keying sequence you used.

_____

How did you find this?   EASY   OK   HARD

18 Writing Fractions

Let's look at ...
- unit fractions
- estimating the fraction of a shape that is shaded
- giving one number as a fraction of another

These are the points you need to know.

✓ **Unit fractions** have a numerator of 1.
  Example $\frac{1}{8}, \frac{1}{12}, \frac{1}{16}$

✓ We can write **one number as a fraction of another**.
  Example 80 as a fraction of 240 is $\frac{80}{240} = \frac{1}{3}$.

Always cancel fractions to their lowest terms.

A *Name that fraction*

Remember to write all fractions in their simplest form.

a What fraction of each shape is shaded?

i _____

ii _____

iii  _____

b Estimate the fraction of the pie chart that is

Car colours produced by a factory

i Red _____
ii Silver _____

Sources of energy in the United States

iii Petroleum _____
iv Other _____

B *Quick questions*

a What fraction of 1 m is 75 cm? _____

b What fraction of £3 is £1·50? _____

c What fraction of 2 kg is 300 g? _____

d What fraction of 4 hours is 30 minutes? _____

e What fraction of 6 hours is 40 minutes? _____

f What fraction of 2 m is 5 cm? _____

g What fraction of 12 is 8? _____

h Give 49 as a fraction of 56. _____

C *Fishing contest*

Asad counted the number of spectators at the finals of a trout fishing contest.
There were 85 men, 20 women and 15 children.

a What fraction were women? _____

b What fraction were men? _____

How many spectators were there altogether?

This bar chart shows the number of trout caught by each of the finalists.

c What fraction of the finalists caught no fish? _____

d What fraction of the finalists caught 2 fish? _____

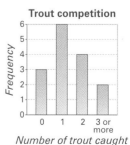

Trout competition

How many finalists were there altogether?

e What fraction of the finalists caught 2 or more fish? _____

How did you find this?   EASY   OK   HARD

19 Fraction and Decimals

Let's look at ...
- writing a decimal as a fraction
- writing a fraction as a decimal

These are the points you need to know.

✓ To write a **decimal as a fraction** write it with a denominator of 10, 100 or 1000. Then cancel to the simplest form.

Example $0.82 = \frac{82}{100}$ $0.785 = \frac{785}{1000}$
$= \frac{41}{50}$ $= \frac{157}{200}$

✓ To write a **fraction as a decimal** we can
1 use known facts   Example $\frac{3}{5} = 3 \times \frac{1}{5} = 3 \times 0.2 = 0.6$
2 make the denominator 10 or 100   Example $\frac{7}{25} = \frac{28}{100} = 0.28$
3 divide the numerator by the denominator   Example $\frac{4}{7} = 4 \div 7 = 0.57$ to 2 d.p.
4 use a calculator   Example $\frac{36}{42}$ key 36 ÷ 42 = to get 0.86 (2 d.p.)
or key 36 $a^{b/c}$ 42 = $a^{b/c}$ to get 0.86 (2 d.p.)

```
7 ) 4·0
    3·5      7 × 0·5
    0·50
    0·49     7 × 0·07
    0·010
    0·007    7 × 0·001
    0·003
Answer  0·571 R 0·003
        0·57 to 2 d.p
```

✓ A **terminating decimal** has a finite number of digits.

✓ A **recurring decimal** has one or more repeating digits.

Example $\frac{2}{3} = 0.6666 ...$ $\frac{5}{11} = 0.454545 ...$
$= 0.\dot{6}$ $= 0.\dot{4}\dot{5}$

(A) Quick changes

1 Write these as fractions in their lowest terms.

a 0·4 _____ **b** 0·35 _____ **c** 0·367 _____ **d** 0·08 _____ **e** 0·850 _____
f 0·600 _____ **g** 0·488 _____ **h** 0·064 _____ **i** 6·2 _____ **j** 3·71 _____

2 Write these as decimals. Do **not** use a calculator.

a $\frac{4}{5}$ _____ **b** $\frac{3}{25}$ _____ **c** $\frac{9}{20}$ _____ **d** $\frac{5}{8}$ _____
e $\frac{9}{30}$ _____ **f** $\frac{29}{50}$ _____ **g** $2\frac{3}{8}$ _____ ***h** $\frac{72}{60}$ _____

3 Use division to write these as decimals. Don't use your calculator. Round to 2 d.p.

a $\frac{2}{9}$ _____ **b** $\frac{19}{3}$ _____ **c** $\frac{3}{7}$ _____

4 Use your calculator to write these as decimals. Round to 2 d.p.

a $\frac{5}{23}$ _____ **b** $\frac{138}{169}$ _____ **c** $\frac{47}{38}$ _____ **d** $\frac{28}{47}$ _____

(B) Odd pair out

Shade the odd pair.   0·3 $\frac{3}{10}$ 0·36 $\frac{9}{25}$ 0·875 $\frac{7}{8}$ 0·18 $\frac{18}{200}$ 1·35 $1\frac{7}{20}$

(C) Sort them

Fractions in their lowest terms with denominators which have prime factors other than 2 and 5 will give recurring decimals.

Decide whether each of these fractions converts to a terminating or recurring decimal.

$\frac{2}{3}$ $\frac{3}{12}$ $\frac{16}{22}$ $\frac{18}{45}$ $\frac{60}{81}$ $\frac{17}{132}$ $\frac{21}{16}$

Terminating

Recurring
$\frac{2}{3}$

*(D) Terminate here

7	11	8
15	5	9
21	3	24
55	4	14

Samantha used two of the numbers in this box to make $\frac{3}{4}$, a fraction which converts to a **terminating decimal**.
Use the rest of the numbers to make five more fractions which convert to **terminating decimals**. You may use each number only once.

How did you find this? EASY OK HARD

20 Ordering Fractions

Let's look at ...
● comparing fractions

✓ We can **compare fractions** by
1 writing them with a common denominator **or**
2 writing them as decimals.

Example Compare $\frac{3}{4}$ and $\frac{5}{7}$.

1 The LCM of 4 and 7 is 28.
So 28 is the common denominator.

Remember the LCM is the lowest common multiple.

$$\frac{3}{4} \xrightarrow{\times 7} = \frac{21}{28} \qquad \frac{5}{7} \xrightarrow{\times 4} = \frac{20}{28}$$

$$\frac{21}{28} > \frac{20}{28}$$
$$\frac{3}{4} > \frac{5}{7}$$

2 $\frac{3}{4} = 0.75$
$\frac{5}{7} = 0.71$ (2 d.p.)
$0.75 > 0.71$
$\frac{3}{4} > \frac{5}{7}$

You can find $\frac{5}{7}$ as a decimal using a calculator or by dividing.

These are the points you need to know.

A Bigger is better

Write these fractions with a common denominator, then circle the biggest.

a $\frac{1}{6} = \boxed{\frac{1}{6}}$ $\boxed{\frac{1}{3} = \frac{2}{6}}$

b $\frac{1}{4} = \boxed{}$ $\frac{3}{8} = \boxed{}$

c $\frac{4}{5} = \boxed{}$ $\frac{7}{10} = \boxed{}$

d $\frac{2}{3} = \boxed{}$ $\frac{3}{4} = \boxed{}$

e $\frac{2}{5} = \boxed{}$ $\frac{3}{8} = \boxed{}$

*f $\frac{13}{15} = \boxed{}$ $\frac{5}{6} = \boxed{}$

Write these fractions as decimals, then circle the biggest.

You may use your calculator.

g $\frac{5}{9} = \boxed{}$ $\frac{10}{17} = \boxed{}$

h $\frac{2}{11} = \boxed{}$ $\frac{3}{19} = \boxed{}$

i $\frac{6}{7} = \boxed{}$ $\frac{19}{21} = \boxed{}$

B Sophie and Simon

a Sophie spent $\frac{2}{5}$ of her birthday money on clothes, and $\frac{3}{7}$ on CDs.
What did she spend more money on? _____

b Simon spent $\frac{3}{8}$ of his holiday in France and $\frac{4}{11}$ in Holland.
In which country did he stay longer? _____

c In a vote for year group representative Sophie received $\frac{6}{13}$ of the votes and Simon received $\frac{4}{9}$ of the votes. Who was elected? _____

C All in order

a Show these fractions with an arrow on this number line.
$\frac{1}{5}, \frac{7}{20}, \frac{6}{15}, \frac{36}{30}, \frac{40}{25}$

b Show these fractions with an arrow on this number line.
$\frac{3}{10}, \frac{9}{25}, \frac{1}{20}, \frac{8}{40}, \frac{6}{15}$

c Put these in order from smallest to largest.
i $\frac{3}{4}, \frac{1}{2}, \frac{5}{8}, \frac{7}{16}$ _____
ii $\frac{2}{5}, \frac{3}{10}, \frac{7}{20}, \frac{1}{2}, \frac{3}{5}$ _____

Try writing each of the fractions with a common denominator.

*D Puzzle time

a I am bigger than $\frac{1}{2}$ but smaller than $\frac{7}{12}$. My denominator is 24. What fraction am I? _____

b I am half way between $\frac{3}{4}$ and $\frac{7}{8}$. I am written in my simplest form. What fraction am I? _____

21 Adding and Subtracting Fractions

Let's look at ...
- adding and subtracting fractions which do not have the same denominator

These are the points you need to know.

✓ To **add and subtract fractions** which do not have the same denominator we find equivalent fractions.

Examples
$$\frac{5}{8} + \frac{2}{3} = \frac{15}{24} + \frac{16}{24}$$
$$= \frac{15 + 16}{24}$$
$$= \frac{31}{24}$$
$$= 1\frac{7}{24}$$

$$\frac{5}{8} - \frac{3}{5} = \frac{25}{40} - \frac{24}{40}$$
$$= \frac{25 - 24}{40}$$
$$= \frac{1}{40}$$

Always write the answers in their simplest form.

✓ We could use the $\boxed{a^{b/c}}$ key on the calculator.

Example $\frac{5}{8} + \frac{2}{3}$ key $\boxed{5}\ \boxed{a^{b/c}}\ \boxed{8}\ \boxed{+}\ \boxed{2}\ \boxed{a^{b/c}}\ \boxed{3}\ \boxed{=}$ to get $\boxed{1\ \lrcorner 7\lrcorner 24.}$
which is $1\frac{7}{24}$.

A Fraction chains

Complete these fraction chains. Write all fractions in their simplest form.

a $\boxed{\frac{1}{2}} \xrightarrow{+\frac{1}{3}} \boxed{\frac{5}{6}} \xrightarrow{-\frac{1}{6}} \boxed{} \xrightarrow{-\frac{1}{12}} \boxed{} \xrightarrow{+\frac{5}{12}} \boxed{}$

b $\boxed{\frac{3}{4}} \xrightarrow{+\frac{1}{8}} \boxed{} \xrightarrow{-\frac{1}{2}} \boxed{} \xrightarrow{-\frac{1}{8}} \boxed{} \xrightarrow{+\frac{2}{5}} \boxed{} \xrightarrow{-\frac{3}{10}} \boxed{}$

c $\boxed{\frac{2}{3}} \xrightarrow{+\frac{1}{5}} \boxed{} \xrightarrow{-\frac{1}{3}} \boxed{} \xrightarrow{-\frac{2}{5}} \boxed{} \xrightarrow{+\frac{7}{30}} \boxed{} \xrightarrow{-\frac{3}{10}} \boxed{} \xrightarrow{+\frac{2}{3}} \boxed{}$

B Perplexing problems

a My car's petrol tank is $\frac{3}{4}$ full when I leave home. At the end of my journey the tank is $\frac{1}{3}$ full. What fraction of the tank of petrol have I used? _____

b Freddy, George and Scott are buying their teacher a gift. Freddy pays $\frac{1}{3}$ of the cost, George pays $\frac{2}{5}$ of the cost. What fraction does Scott need to pay? _____

C Magic squares

Complete these magic squares so that every row, every column and every diagonal adds up to the same number.

Write the fractions in **b** and **c** in their simplest form.

a

$\frac{2}{8}$		$\frac{5}{8}$
		$\frac{1}{8}$
$\frac{3}{8}$		$\frac{6}{8}$

b

$\frac{2}{5}$		
$\frac{9}{10}$		$\frac{3}{10}$
$\frac{1}{2}$		$\frac{4}{5}$

c

$\frac{1}{6}$	$\frac{7}{12}$	$\frac{1}{2}$
	$\frac{1}{4}$	$\frac{2}{3}$

*D Make it half

$\boxed{\frac{1}{2} = \frac{5}{6} - \frac{1}{3}}$ $\boxed{\frac{1}{2} = \frac{5}{12} + \frac{2}{24}}$ $\boxed{\frac{1}{2} = \frac{1}{3} - \frac{1}{4} + \frac{1}{5} + \frac{13}{60}}$

Make $\frac{1}{2}$ in as many ways as you can.
You may add or subtract.
You may use as many fractions as you like in each calculation.
In each calculation, all the fractions must have a **different denominator**.

How did you find this? EASY OK HARD

22 Multiplying and Dividing Integers by Fractions

Let's look at ...
● finding a fraction of an integer
● dividing an integer by a fraction

These are the points you need to know.

✓ $\frac{1}{8}$ of 24, $\frac{1}{8} \times 24$, $24 \times \frac{1}{8}$, $24 \div 8$ are all equivalent.
We can **multiply integers and fractions**.

In maths, 'of' means multiply.

Examples
$$3 \times \frac{5}{8} = 3 \times 5 \times \frac{1}{8}$$
$$= \frac{15}{8}$$
$$= 1\frac{7}{8}$$

$$\frac{2}{3} \text{ of } 7 = 2 \times \frac{1}{3} \times 7$$
$$= \frac{14}{3}$$
$$= 4\frac{2}{3}$$

$$1\frac{2}{3} \times 12 = \frac{5}{3^1} \times \frac{12^4}{1}$$
$$= \frac{20}{1}$$
$$= 20$$

We have cancelled the 12 and the 3.

✓ $8 \div \frac{1}{4}$ means 'How many $\frac{1}{4}$s are there in 8?'
The answer is 32.

A Puzzle

$$\frac{1}{\square} \times 3\square = 1\square$$

Each square represents 1 digit.

The puzzle in the box has nine possible solutions.

One solution is $\frac{1}{2} \times 3\boxed{0} = 1\boxed{5}$

Find the other eight solutions. _____ _____ _____

_____ _____ _____ _____

B Dividing with diagrams

Use the diagrams to help find the answers to these.

a How many fifths are there in 4 cakes? _____

b How many eighths are there in 3 pizzas? _____

*c $4 \div \frac{2}{3} =$ _____

Colour lots of $\frac{2}{3}$.

*d $8 \div \frac{4}{7} =$ _____

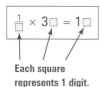

Divide the line into lots of $\frac{4}{7}$.

C Torture chambers

| $\overline{35}$ | $\overline{12}$ | $\overline{40}$ | | $\overline{28}$ | $\overline{6\frac{2}{3}}$ | $\overline{7\frac{6}{7}}$ | $\overline{77}$ | $\overline{6\frac{2}{3}}$ | $\overline{24}$ | $\overline{77}$ | $\overline{7\frac{1}{5}}$ | | $\overline{48}$ | $\overline{9}$ | | $\overline{12}$ |

| | | | | **L** | | | | | | | | | | | | | |
| $\overline{24}$ | $\overline{6}$ | $\overline{77}$ | | $\overline{77}$ | $\overline{\mathbf{20}}$ | $\overline{77}$ | $\overline{150}$ | $\overline{24}$ | $\overline{6\frac{3}{4}}$ | $\overline{28}$ | $\overline{150}$ | | $\overline{150}$ | $\overline{6}$ | $\overline{12}$ | $\overline{28}$ | $\overline{6\frac{3}{4}}$ |

| $\overline{7\frac{1}{5}}$ | $\overline{77}$ | $\overline{6\frac{2}{3}}$ | $\overline{24}$ | $\overline{28}$ | $\overline{40}$ | $\overline{24}$ |

Write the letter beside each question above its answer in the box.

Write **N**, **R**, **D** and **V** as mixed numbers.

L $\frac{4}{5}$ of 25 = **20** **H** $\frac{2}{7}$ of 21 **Y** $\frac{3}{8} \times 24$ **A** $\frac{4}{9}$ of 27 **S** $\frac{5}{6} \times 48$

T $\frac{3}{10}$ of 80 **C** $\frac{5}{6}$ of 180 **W** $1\frac{3}{4}$ of 20 **E** $2\frac{1}{5}$ of 35 **N** $\frac{2}{3}$ of 10

R $\frac{3}{4}$ of 9 **D** $\frac{2}{5} \times 18$ **V** $\frac{5}{7}$ of 11 **I** $7 \div \frac{1}{4}$ **B** $6 \div \frac{1}{8}$

How did you find this? **EASY** **OK** **HARD**

23 Fractions, Decimals and Percentages

Let's look at ...
● converting fractions and decimals to percentages

✓ We can write **fractions and decimals as percentages** by
  ● writing with denominator 100 **or**
  ● multiplying by 100%

Work this out on the calculator

These are the points you need to know.

Examples $\frac{3}{25} = \frac{12}{100}$ $\frac{3}{14} = (3 \div 14) \times 100\%$ $1 \cdot 27 = 1 \cdot 27 \times 100\%$
          $= 12\%$ $= 21 \cdot 4\%$ to 1 d.p. $= 127\%$

✓ Some common fractions are written as mixed number percentages.
  $\frac{1}{3} = 33\frac{1}{3}\%$ $\frac{2}{3} = 66\frac{2}{3}\%$ $\frac{1}{8} = 12\frac{1}{2}\%$

A Table time

Complete these tables. Do not use your calculator.

Fraction	Decimal	Percentage
$\frac{4}{5}$		
	0·6	
		40%
		45%
	0·03	

Fraction	Decimal	Percentage
$\frac{3}{25}$		
$\frac{21}{30}$		
		172%
$2\frac{13}{20}$		
	0·3̇	

B Quick changes

Change these to percentages.

a 0·42 _____ **b** 0·63 _____ **c** 1·54 _____ **d** 0·585 _____

Write these as percentages to 1 d.p. You may use your calculator.

 e $\frac{6}{13}$ _____ **f** $\frac{16}{35}$ _____ **g** $\frac{121}{183}$ _____ **h** $\frac{412}{61}$ _____

C Match it up

Choose a percentage from the circle to match each sentence.

20% $66\frac{2}{3}\%$ 5% 60% 70% 19% 36% $87\frac{1}{2}\%$ 95% 40%

a One part in five of a ripe banana is sugar. _____
b Two out of three people in a town eat breakfast. _____
c 12 out of 30 students in a class own a mobile phone. _____
d 19 parts out of 20 in a tomato are water. _____
e Seven out of eight people at a tennis match wore hats. _____

D Susan's collection

Susan collects rocks. She estimates the proportion of some minerals in some of her rocks.

		Minerals	
		Quartz	Feldspar	Mica
Rocks	Gneiss	$\frac{3}{8}$	$\frac{1}{2}$	$\frac{1}{8}$
	Schist	0·22	0·66	0·12
	Slate	$\frac{1}{4}$	$\frac{3}{10}$	$\frac{1}{20}$

a What percentage of slate is mica? _____
b Which rock has the least quartz in it? _____
  What percentage is this? _____
c Which rock has the most mica in it? _____
  What percentage is this? _____
*d One rock has quartz, feldspar, mica and a **fourth mineral** in it. Which rock is this? _____
  What is the percentage of the fourth mineral? _____

How did you find this? [EASY] [OK] [HARD]

24 Finding Percentages

Let's look at ...
● finding 'percentages of' mentally, using written methods and using a calculator

These are the points you need to know.

✓ We can find **'percentage of'** mentally.

Example   Find 60% of £480.
50% of £480 = £240   10% of £480 = £48
so 60% of £480 = £240 + £48 = £288

✓ We can find **'percentage of'** using a written method.

Example
Find 16% of 78.

Using fractions

$16\% \text{ of } 78 = \frac{16}{100} \times 78$

$\begin{array}{r} 78 \\ \times\ 16 \\ \hline 780 \\ 468 \\ \hline 1248 \end{array}$

$= \frac{16 \times 78}{100}$
$= \frac{1248}{100}$
$= \mathbf{12 \cdot 48}$

Using decimals

$16\% \text{ of } 78 = 0 \cdot 16 \times 78$
$= 16 \times 78 \div 100$
$= 1248 \div 100$
$= \mathbf{12 \cdot 48}$

Finding 1% first

$1\% \text{ of } 78 = 0 \cdot 78$
$16\% \text{ of } 78 = 16 \times 0 \cdot 78$
$= 16 \times 78 \div 100$
$= 1248 \div 100$
$= \mathbf{12 \cdot 48}$

✓ We can find **'percentage of'** using a calculator.

Example
Find 17% of 84.

Using fractions

$17\% \text{ of } 84 = \frac{17}{100} \times 84$
Key ⟨17⟩⟨÷⟩⟨100⟩⟨×⟩⟨84⟩⟨=⟩
to get **14·28**.

Using decimals

$17\% \text{ of } 84 = 0 \cdot 17 \times 84$
Key ⟨0·17⟩⟨×⟩⟨84⟩⟨=⟩
to get **14·28**.

Finding 1% first

$1\% \text{ of } 84 = 0 \cdot 84$
$17\% \text{ of } 84 = 17 \times 0 \cdot 84$
Key ⟨17⟩⟨×⟩⟨0·84⟩ to get **14·28**.

Ⓐ *Howzat?*

Find the answers to each of the following **mentally**, then shade them in the grid.
The largest two numbers left at the end are the minimum and maximum number of stitches used to hold a cricket ball together.

72	96	123	74
11	65	120	94
121	44·4	14	27·9
70	104	150	152

a 20% of 70 = ____
b 5% of 220 = ____
c 60% of 120 = ____
d 95% of 160 = ____
e $33\frac{1}{3}$% of 369 = ____
f 110% of 110 = ____
g 65% of 160 = ____
h 47% of 200 = ____
i 31% of 90 = ____

A cricket ball has between ____ and ____ stitches.

Ⓑ *Show your skills*

Use a written method to find each of these. Show your working.

a | 18% of 80 cm
b | 72% of £80
c | 43% of 62 kg
d | $17\frac{1}{2}$% of £540

*Ⓒ *Moving time*

a The Reillys buy their new house for £236 000. The real estate agent gets 4% of this amount. How much does the agent get? _____

b The Reillys can only afford to lay carpet in 65% of their house. If the floor area is 186 m^2, what will the area of the carpet be? _____

c The garage takes up 6·5% of their 512 m^2 lot. What is the floor area of the garage? _____

How did you find this?   ⟨ EASY ⟩ ⟨ OK ⟩ ⟨ HARD ⟩

25 Percentage Increase and Decrease

> **Let's look at ...**
> ● finding the outcome of a given percentage increase or decrease

These are the points you need to know.

✓ To find an **increase of 5%** we can find 5% and add it on **or** we can find 105%.

✓ To find a **decrease of 18%** we can find 18% and subtract it **or** we can find 82%. 82% = 100% − 18%.

Example Increase 80 by 45%.
45% of 80 = 36   **or**   145% of 80 = 1·45 × 80
Answer = 80 + 36   = **116**
= **116**

Example Declan bought a stereo for £225. He sold it for a 23% loss.
23% of £235 = £51·75   **or**   77% of £225 = 0·77 × 225
£225 − £51·75 = £173·25   = **£173·25**
He sold it for **£173·25**.

(A) Store wide sale

Calculate the sale price of these items.

a
Was £32
20% off
Sale price = ____

b
Were £90
15% off
Sale price = ____

c
Perfume
Was £42
40% off
Sale price = ____

d
Was £9.50
24% off
Sale price = ____

(B) Antique auction

Isla bought this antique jewellery at an auction.

 £35   £105   £60

Two years later she took it in to be valued.

a The earrings increased in value by 100%. Isla said, 'They are now worth £35.'
Is she right or wrong? _____ Explain. _____

b The necklace increased in value by 300%. Isla said, 'it is now worth £315.'
Is she right or wrong? _____ Explain. _____

c Isla sold the brooch for 30% more than she paid for it.
Which of these should you multiply £60 by to find out how much she sold it for?
A 0·3   **B** 1·03   **C** 1·3   **D** 0·03   _____

(C) Patrick's car

On Monday Patrick advertises his car for £1200.
He decides to reduce the price by 5% every day until it sells.
Sean would like to buy the car, but he only has £1050.

a On which day could he buy it? _____   **b** How much change would he get? _____

(D) Winter retreat

Three hotels on a resort island decide to increase their daily rate by 15%.
Calculate the new daily rate for each hotel.

a Beachcomber Hotel, was £120 per day. _____

b Sunrise Inn, was £132 per day. _____

c Palm Tree Lodge, was £75 per day. _____

How did you find this?   EASY   OK   HARD

26 Direct Proportion

Let's look at ...
● solving problems that involve direct proportion

These are the points you need to know.

✓ Sometimes we use **proportional reasoning** to solve problems.

Example

+1(1 tie costs £7·25) + 7·25
+1(2 ties cost £14·50) + 7·25
    3 ties cost £21·75

As the number of ties increases so does the cost.

The number of ties and the cost of the ties are in **direct proportion**.

✓ *Example* 9 mini-bites cost £5·22. To find the cost of 5 mini-bites, first find the cost of 1.
1 mini-bite costs $\frac{£5·22}{9}$.
5 mini-bites cost $\frac{5·22}{9} \times 5 =$ **£2·90**.

A McGregor's Garden Shop

Use the price list to calculate the costs of Eliza and Carl's shopping lists.

McGregor's Price List
6 roses cost £18
5 camelias cost £22·50
12 petunias cost £7·80
9 lavenders cost £24·75
8 gnomes cost £100
7 statues cost £320·95

Eliza's Shopping List	
8 roses	_____
3 camelias	_____
14 petunias	_____
7 lavenders	_____
5 gnomes	_____
12 statues	_____
Total cost	_____

Carl's Shopping List	
5 roses	_____
7 camelias	_____
3 petunias	_____
2 lavenders	_____
9 gnomes	_____
14 statues	_____
Total cost	_____

B Mix it up

a A garden fertiliser is made by mixing 40 g of crystals in 5 ℓ of water.
What mass of crystals should be mixed with 3 ℓ of water? _____

b Karen usually mixes 20 g of milk powder with 150 mℓ of water to make her baby's formula drink.
Today she uses 210 mℓ of water. How much powder does she need? _____

c Callum usually makes gravy by mixing 50 g of gravy powder with 1 ℓ of water.
Today he only has 20 g of gravy powder left. How much water should he add? _____

*C Birthday treat

8 magazines cost £36

4 chocolates cost £2.60

Vanessa wants to buy 5 or 6 magazines and some chocolates for her mother.
She wants to spend exactly £40.
How many chocolates should she buy? _____

27 Simplifying Ratios

Let's look at ...
● writing ratios in their simplest form

These are the points you need to know.

✓ We always simplify **a ratio to its simplest form**. We do this by cancelling.

Example   6 : 9 : 12
÷3 (÷3) ÷3   **All parts have been divided by 3.**
= 2 : 3 : 4

✓ A ratio in its simplest form does **not have fractions or decimals**.

Example $3\frac{1}{2}$: 5 = 7 : 10   **Both parts have been multiplied by 2 to get whole numbers.**

✓ All parts of a ratio must be in the **same units**.

Example   1 mm : 5 m = 1 mm : 5000 mm   45c : \$2 = 45c : 200c
                        = 1 : 5000   = 45 : 200
                                                          = 9 : 40

A Spot the problem

Describe the problem with each of these ratios, and write the ratio correctly.

a   The ratio of cows to sheep at a farm is 12 : 40. _____
_____

b   The ratio of orange juice to apple juice in a drink is 2·5 ℓ : 2 ℓ. _____
_____

c   The ratio of adult entry fee to child entry fee to visit a museum is £1·90 : 80p. _____
_____

B Quick questions

Write these ratios in their simplest form.

a   4 : 12 _____   **b**   10 : 25 _____   **c**   6 : 9 : 12 _____

d   350 : 600 : 250 _____   **e** $3\frac{1}{2}$: 2 _____   **f**   4·5 : 4 _____

g   2·1 : 4 _____   **h**   1 : $3\frac{1}{3}$ _____   **i**   3 cm : 5 cm _____

j   13p : £2 _____   **k**   20 cm : 6 m _____   **l**   1 kg : 460 g _____

C Work wonderings

a   Arshad works at a newsagent for 27 hours per week, and at a garage for 15 hours per week. Write, in its simplest form, the ratio of newsagent hours to garage hours. _____

*****b**   This week Philippa picked 16·5 kg of apples. Last week she picked 7 kg less. Write the ratio of apples picked last week to apples picked this week. _____

*D Puzzle time

a   The length and width of a rectangle are in the ratio 2 : 3.
The area of the rectangle is 24 cm².
What are the length and width of the rectangle? _____ , _____

b   The length and width of another rectangle are in the ratio 7 : 10.
The perimeter of the rectangle is 68 mm.
What are the length and width of the rectangle? _____ , _____

How did you find this?   EASY   OK   HARD

28 Ratio and Proportion

Let's look at ...
● writing ratios and proportions

These are the points you need to know.

✓ **Proportion** compares part to whole.
 Ratio compares part to part.
  Example The ratio of cherry to carrot to
  chocolate cakes = 6 : 24 : 20
      = 3 : 12 : 10
  The proportion of carrot cakes = $\frac{24}{50}$
      = $\frac{12}{25}$

Cakes	Number
Cherry	6
Carrot	24
Chocolate	20

There are 6 + 24 + 20 = 50 cakes altogether.

A Pay your rent

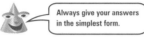

Always give your answers in the simplest form.

This table shows the rent paid by three flatmates.

Name	Rent paid
Tony	£80
Denise	£60
Ian	£50

a Write these as ratios.
 i Tony's to Denise's rent. _____
 ii Denise's to Ian's to Tony's rent. _____
b What proportion of the rent does Tony pay? _____
c What proportion of the rent does Ian pay? _____
d What fraction of Tony's rent does Denise pay? _____

B Morning Muesli

A 1 kg bag of Morning Muesli contains the ingredients given in the table.

a Write these as ratios.
 i Oats to barley. _____
 ii Nuts to dried fruit. _____
 iii Barley to nuts to oats. _____

Ingredient	Quantity (g)
Oats	400
Nuts	150
Barley	250
Dried fruit	200

b What proportion of the muesli was
 i Oats _____ **ii** Nuts _____ **iii** Dried fruit _____

 Give these proportions as decimals.

C Cross country

In a team race, Henry and his friends run the distances given in this table.

Name	Distance
Henry	9 km
Edward	11 km
Mary	12 km
Jane	18 km

a Write the ratio of Henry's to Jane's distance. _____
b Write the ratio of Mary's to Edward's to Jane's distance. _____
c Give the proportion of the race Edward ran as a fraction. _____
d Give the proportion of the race Henry ran as a percentage. _____
e What proportion of the race did the girls run?
 Give your answer as a percentage. _____

29 Solving Ratio and Proportion Problems

Let's look at ...
● real-life problems involving ratio and proportion

✓ We can solve real-life **ratio and proportion problems**.

Example   A nut mix has brazils, cashews and almonds in the ratio 2 : 3 : 5.
We can calculate how many grams of brazils and cashews are needed to mix with 350 g of almonds.

5 parts = 350 g
1 part   = 70 g   **350 ÷ 5**
2 parts = 140 g   **2 × 70 g**
3 parts = 210 g   **3 × 70 g**

We need **140 g of brazils** and **210 g of cashews** to mix with 350 g of almonds.

(A) Devonshire teas

This recipe makes 15 scones.
How much of each of these is needed to make 40 scones?

a   milk _____

b   butter _____

c   flour _____

d   baking powder _____

Scones
3 cups flour
$4\frac{1}{2}$ tsp baking powder
$\frac{1}{2}$ tsp salt
45 g butter
270 mℓ milk

(B) How far?

The scale on a map is 1 : 2000.
Candice measured these distances on the map.
How far are they in real life? Give the answers in metres.

a   1 cm _____

b   6 cm _____

c   12·1 cm _____

d   19·2 cm _____

(C) Cat care

A vet clinic treats 300 cats in a year.
How many were vaccinated? _____

Cats vaccinated

Remember there are 360° in a full circle.

(D) Shady figures

A 15 metre high tree casts a 12 metre long shadow. At the same time

a   a flagpole casts a 24 m shadow. How tall is the flagpole? _____

✱**b**   a building casts a 30 m shadow. How tall is the building? _____

✱(E) Perplexing puzzles

a   The ratio of males to females in a yacht race is 4 : 3.
If 5 males drop out of the race there will be the same number of males and females.
How many people are in the race now? _____

b   The ratio of Stephanie's age to her son Matthew's age is 5 : 1.
Stephanie is older than 18 and younger than 42.
In two years the ratio of Stephanie's age to Matthew's age will be 4 : 1.
How old is Matthew now? _____

How did you find this?   EASY   OK   HARD

30 Dividing in a Given Ratio

Let's look at ...
● dividing a quantity into two or more parts in a given ratio

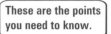

These are the points you need to know.

✓ We can **divide in a given ratio**.

Example   Divide £5650 in the ratio 2 : 3 : 5.

Add the parts of the ratio.

There are 2 + 3 + 5 = 10 shares altogether.

1 share or $\frac{1}{10}$ is $\frac{5650}{10}$ = £565

2 shares or $\frac{2}{10}$ is 2 × £565 = £1130

3 shares or $\frac{3}{10}$ is 3 × £565 = £1695

5 shares or $\frac{5}{10}$ is 5 × £565 = £2825

Check that these add up to £5650.

£5650 divided in the ratio 2 : 3 : 5 is **£1130 : £1695 : £2825**.

(A) Winnings

a Debbie, Edward and Frank bought a £5 lottery ticket. Debbie paid £1, Edward and Frank each paid £2.
How much should Frank get if their winnings are

i £60 _____   ii £750 _____   iii £1240 _____   iv £198·50? _____

b Kamil, Giles and Lily paid for a lottery ticket in the ratio 3 : 5 : 6.
How much should Giles get if their winnings are

i £140 _____   ii £392 _____   iii £938 _____   iv £11 417? _____

(B) Mix it up

a A rose fertiliser contains 4 parts nitrogen, 5 parts phosphate and 9 parts potassium.
How much of each is in a 900 g bag?

Nitrogen _____   Phosphate _____   Potassium _____

b A tomato soup is made from 4 parts soup concentrate, 1 part water and 2 parts milk.
How much soup concentrate is required to make 3·5 ℓ of soup? _____

(C) What's the angle?

a Two angles on a straight line are in the ratio 1 : 5.
What is the size of each angle? _____, _____

b The three angles in a triangle are in the ratio 2 : 3 : 7.
What is the size of each angle? _____, _____, _____

*c The four angles in a trapezium are in the ratio 2 : 3 : 6 : 7.
What is the size of each angle? _____, _____, _____, _____

*(D) Kings versus Queens

Each year King's College and Queen's College hold a winter sports tournament.

Number of games played	
Rugby	– 16
Football	– 27
Hockey	– 18
Netball	– 20

Ratio of games won by King's to games won by Queen's			
Rugby	3 : 5	Hockey	4 : 5
Football	2 : 1	Netball	3 : 2

Which school was the overall winner for this year? _____

By how many games did they beat the other school? _____

ALGEBRA

31 Getting Started with Algebra

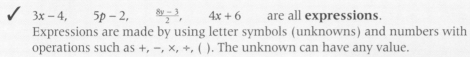

Let's look at ...
- expressions, equations, formulae and functions
- rules for writing expressions

✓ $3x - 4$, $5p - 2$, $\frac{8y-3}{2}$, $4x + 6$ are all **expressions**.
Expressions are made by using letter symbols (unknowns) and numbers with operations such as +, −, ×, ÷, ( ). The unknown can have any value.

✓ $2b + 7 = 16$ is an **equation**, b has a particular value.
$n + m = 12$ is also an **equation**, n and m can have any values as long as they add to 12.

> Equations and formulae have an equals sign. Expressions do not.

> These are the points you need to know.

✓ $C = \frac{3}{5}m$ is a **formula**. C is the sale price of any item which originally cost £m.
If we know m, we can work out C.
In a formula the letters stand for something specific.

> A formula always has more than one unknown.

✓ $y = 3x - 2$ is a **function**. We can work out y for any given value of x.

> Functions are a special sort of equation.

✓ **Expressions** are usually written without a multiplication or division sign.

Examples $2 \times n$ is written as $2n$. ← We write the number first.
$1 \times n$ or $n \times 1$ is written as n.
$p \times q$ or $q \times p$ is written as pq. ← We usually write the letters in alphabetical order.

$(a + b) \div c$ is written as $\frac{a+b}{c}$.
$x \times (y + 4)$ is written as $x(y + 4)$.
$n \times n$ is written as n^2.

(A) Which is which?

In each of these rows there is *one* equation, *one* expression, *one* formula and *one* function. Copy each into the correct box below.

a $c + d = 8$ $y = 2x + 1$ $3x - 1$ $W = 6D$ (W = wages, D = days)

b $\frac{6m-3}{2}$ $s = 5t^2$ (s = speed, t = time) $y = \frac{4x-1}{3}$ $5x - 1 = 7$

c $y = \frac{x}{8} + 6$ $v = l^3$ (v = volume, l = length) $4a + c$ $6e + f = 12$

d $D = \frac{m}{v}$ (D = density, m = mass, v = volume) $y = \frac{x+4}{5}$ $\frac{2p-q}{3} = 7$ $\frac{7e-d}{8}$

Expressions	Equations	Formulae	Functions

(B) Why is that?

a Is $3x - y$ an expression or an equation? _____ How can you tell? _____
_____ Can x and y have any values? _____

b The price, p, of c chocolate bars is given by $p = 30c$.
Is $p = 30c$ an equation or a formula? _____ How do you know? _____
_____

(C) Find the best

Choose the **best** expression from the oval to match each of these.

a $2 \times a$ _____
b $b \times a$ _____
c $a \times 1$ _____
d $3 \times (a + b)$ _____
e $2 \times a \times a$ _____
f $(2a + b) \div 2$ _____
g $5 \times a \times b$ _____
h three times a, plus b _____
i two times the product of b and a _____

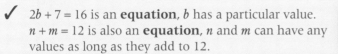

$2a$ a $2ba$
ba $2a^2$ ab $3ab$
$\frac{2a+b}{2}$ $\frac{2(a+b)}{2}$ $2 \times a^2$
$1a$ $5(a+b)$
$3(a+b)$ $5ab$ $3a+b$
$2ab$

---

32 Understanding Algebra

Let's look at ...
- **rules for algebraic operations**
- **inverse operations**

These are the points you need to know.

✓ **Algebraic operations** follow the same rules as **arithmetic operations**.
Order of operations. Algebraic operations are done in this order
Brackets, **I**ndices, **D**ivision and **M**ultiplication, **A**ddition and **S**ubtraction (BIDMAS)

Examples   In $6p - 4$ $6 \times p$ is worked out first.
In $4a^2 + 7$ a^2 is worked out first.

Commutative rule

Examples   If $x + y = z$ then $y + x = z$.   If $mn = p$ then $nm = p$.

Inverse operations

Examples   If $f + g = h$ then $h - g = f$ and $h - f = g$.   If $pq = r$ then $\frac{r}{p} = q$ and $\frac{r}{q} = p$.
If $5x - 6 = y$ then $x = \frac{y + 6}{5}$.

A Circle it

Find the value of each expression if $x = 3$. Circle the part of the expression you worked out first. The first one is done for you.

We multiply before subtracting.

a $14 - \textcircled{3x} = \underline{\quad 5 \quad}$ **b** $5(x - 1) = \underline{\quad\quad}$ **c** $\frac{15(x - 2)}{3} = \underline{\quad\quad}$ **d** $8 + 4x^2 = \underline{\quad\quad}$

B Whoops!

Explain what is wrong with each of these.

a | $abc = a + b + c$ | _____

b | $105 - 28 = 105 - 30$ |
         | $\quad\quad\quad = 75 + 2$ | _____
         | $\quad\quad\quad = 77$ | _____

C Piece of cake

A bakery has these pieces of cake left at the end of the day. The size of each piece of cake is given by the letter below it.

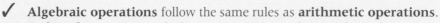

a   b   c   d

a   Decide whether each of these equations are true (T) or false (F).

i $a = b + d$ ____   **ii** $2d + c = a$ ____   **iii** $3b = 2a$ ____   **iv** $\frac{b}{3} = c$ ____   **v** $a - 2d = c$ ____

b   Complete each of these equations in two different ways.

i $b = \underline{\quad\quad}$ **ii** $d = \underline{\quad\quad}$ **iii** $2a = \underline{\quad\quad}$ **iv** $3c = \underline{\quad\quad}$
$b = \underline{\quad\quad}$ $d = \underline{\quad\quad}$ $2a = \underline{\quad\quad}$ $3c = \underline{\quad\quad}$

D Shade it

You could check by choosing numbers to substitute for the unknowns.

Shade two matching equations from the box for each of these. One has been done for you.

a $c + d = 8$

| $8 - d = c$ | $c = d + 8$ | $d = 8 - c$ | $d - c = 8$ | $d = \frac{8}{c}$ |

b $3p = q$

| $3q = p$ | $\frac{q}{3} = p$ | $3 = \frac{p}{q}$ | $3 = \frac{q}{p}$ | $pq = 3$ |

c $\frac{m}{4} = n$

| $4 = nm$ | $4 = \frac{m}{n}$ | $m = n - 4$ | $n + 4 = m$ | $4n = m$ |

Shade one matching equation for each of these.

*d $4x - 1 = y$

| $\frac{y}{4} + 1 = x$ | $\frac{x}{4} + 1 = y$ | $\frac{y + 1}{4} = x$ |

*e $\frac{a + 2}{3} = b$

| $a = 3(b - 2)$ | $a = 3b - 2$ | $a = 2b - 3$ |

How did you find this?   **EASY**   **OK**   **HARD**

33 Simplifying Expressions

Let's look at ...
● simplifying multiplication expressions
● simplifying division expressions by cancelling

✓ We can **simplify expressions**.

Examples **Multiplication**

$$3a \times 2a = 3 \times a \times 2 \times a$$
$$= 3 \times 2 \times a \times a$$
$$= 6a^2$$

$$n^2 \times n^4 = n \times n \times n \times n \times n \times n$$
$$= n^6$$

We can cancel by dividing both numerator and denominator by the same thing.

Division

$$\frac{18p}{12} = \frac{\cancel{18}^3 p}{\cancel{12}_2} = \frac{3p}{2}$$

$$\frac{y^4}{y^2} = \frac{\cancel{y} \times \cancel{y} \times y \times y}{\cancel{y} \times \cancel{y}} = y^2$$

These are the points you need to know.

A Did you know?

					A						A	
$3y^2$	$10y^3$	$30y$	$10y^3$	$30y$	**16y**	$12y^2$	$3y^2$	^-y	$12y$		**16y**	
					A							
^-14y	y^3	y^4	$30y$	$4y$	$10y^3$	**16y**	y^3	$30y$	$24y$	$\frac{6y}{5}$	$3y^2$	$10y^3$ $30y$
					A			A		A		
$\frac{6y}{5}$	$24y$	$5y$	$30y$	^-y	$12y$	**16y**	$\frac{6y}{5}$	y^2	**16y**	y^3	y^3	$\frac{4y^2}{3}$ **16y** $12y^2$

Simplify these expressions. Write the letter beside each expression above its answer in the box.

A $8 \times 2y = \mathbf{16y}$ **F** $4 \times 3y$ **E** $5y \times 6$ **U** $y \times y \times y \times y$

L $y \times y \times y$ **T** $3 \times y^2$ **R** $6y \times 2y$ **B** $^-2y \times 7$

I $^-8 \times ^-3y$ **Z** $\frac{15y}{3}$ **M** $y^5 \div y^3$ **O** $\frac{y^3}{y^2}$

***S** $\frac{150y}{125}$ ***H** $2y^2 \times 5y$ ***W** $\frac{20y^2}{5y}$ ***C** $\frac{16y^3}{12y}$

HELP

Need help with multiplying with negative numbers? Go to page 6

B Pyramids

The expression in each circle is found by multiplying the expressions in the two circles below it. Each expression must be simplified.

a

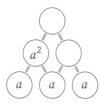

a^2

a a a

b

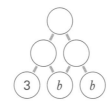

3 b b

c

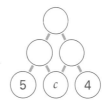

5 c 4

* C Odd one out

Shade the odd one out in each stack.

a

$2 \times 2m$
$\frac{8m}{2}$
$4 \times m^2$
$\frac{24m}{6}$

b

$c \times c \times c$
$c^9 \div c^3$
$c^5 \div c^2$
$\frac{c^6}{c^3}$

c

$a^2 \times 6$
$\frac{12a^2}{2}$
$\frac{18a}{3}$
$3a \times 2a$

d

$h \times 5h$
$\frac{15h^2}{3}$
$\frac{20h^2}{4h}$
$\frac{10h^3}{2h}$

How did you find this?   **EASY**   **OK**   **HARD**

34 Brackets

Let's look at ...
● multiplying out brackets

These are the points you need to know.

✓ When we **multiply out** brackets we use the **distributive law**.

Examples $5(x - 7) = 5 \times x + 5 \times {}^-7$ $3(2a + 4b) = 3 \times 2a + 3 \times 4b$
$= 5x - 35$ $= 6a + 12b$

	x	$^-7$
5	$5x$	$^-35$

	$2a$	$4b$
3	$6a$	$12b$

Ⓐ Quick questions

Multiply out the brackets.

a $5(c + 2) =$ _____

b $3(b - 8) =$ _____

c $2(a - 12) =$ _____

d $6(p + 3) =$ _____

e $4(s - 5) =$ _____

f $3(a + 1) =$ _____

g $2(y - 6) =$ _____

h $^-2(d + 8) =$ _____

i $^-4(c - 4) =$ _____

j $a(b - c) =$ _____

k $5(3g + h) =$ _____

l $11(3y - 10z) =$ _____

Ⓑ Find the mistakes

Sally multiplied out these brackets, but she made some mistakes.
Mark her answers with a ✓ or a ✗.

a $7(c + 3) = 7c + 21$ ☐

b $2(h - 14) = 2h - 14$ ☐

c $6(e + 3) = 6e - 18$ ☐

d $2(a - 8) = {}^-2a + 16$ ☐

e $4(a + 4) = 4a + 16$ ☐

f $5(6a + 3) = 30a + 15$ ☐

g $2(x + y) = x^2 + y^2$ ☐

h $m(n - p) = mn + mp$ ☐

i $3(y + 2z) = 3y + 6z$ ☐

j $4(2q - 3r) = 8q - 12r$ ☐

k $4(h - 6g) = 24h - 24g$ ☐

l $12(5r - 3t) = 60r - 36t$ ☐

Ⓒ Sets of three

$\boxed{2n}$, $\boxed{^-10}$ and $\boxed{2(n - 5)}$ are a set of three because $2(n - 5) = 2n - 10$.

Circle the other sets of three in this grid.

There will be one expression left at the end.

$2(n-5)$	$2n$	$7n$	$9n$	$3(6n-2)$	$18n$	$^-8$	$24n$
$^-10$	$7n$	$7(n-2)$	$^-10$	^-15n	$^-6$	$4(6n-2)$	15
7	$7(n+1)$	$^-14$	^-9n	$^-5(3n+2)$	$^-50$	$21n$	$3(7n+5)$
$12n$	$4(3n+2)$	8	$^-9(n-3)$	$^-27$	$25(3n-2)$	$75n$	75
$2(6n-4)$	$12n$	$^-8$	27	$3(2n-9)$	$6n$	$15(3n+5)$	$45n$

The expression left is _____ .

How did you find this?   [EASY] [OK] [HARD]

35 Collecting Like Terms

Let's look at ...
● simplifying expressions by collecting like terms

These are the points you need to know.

✓ We can simplify expressions by **collecting like terms**.

Examples $8p + 4q - 6p - 7q = 8p - 6p + 4q - 7q$
$$= \mathbf{2p - 3q}$$

$15 - 2(x + 3y) = \mathbf{15 - 2x - 6y}$ because $^-2 \times 3y = ^-6y$

$8x^2 - 3x + 5x^2 + 2x = 8x^2 + 5x^2 - 3x + 2x$
$$= \mathbf{13x^2 - x}$$

We cannot subtract x from $3x^2$ because they are not like terms.

(A) Missing terms

Fill in the missing terms and expressions.

a $4n + 3n + 5x - 2x = 7n + \boxed{}$

b $8a - 4a + 7b - 5b = \boxed{} + \boxed{}$

c $9x - 3a - 6x + 5a = \boxed{} + \boxed{}$

d $12b - 5c - 7b - 2c = \boxed{}$

e $4c + 4 + 2c - 2 + c = \boxed{}$

f $9y + 2 - 3y + 6 - y = \boxed{}$

g $5x^2 + 2x^2 + x^2 = \boxed{}$

h $3x^2 + 2x + 4x^2 = \boxed{}$

i $6n + 2x + \boxed{} + \boxed{} = 8n + 5x$

j $5c + 4d - \boxed{} - \boxed{} = 3c - 6d$

k $6a - \boxed{} + \boxed{} + 8 = a + 11$

l $3x^2 + \boxed{} + \boxed{} + 2x = 6x^2 + 7x$

m $2(3a - 2) + 4(2 - a) = \boxed{}$

n $12 + 3(x + 2) = \boxed{}$

o $4(n + 3) - 2(n + 5) = \boxed{}$

p $6(3n + a) - (4n + 2a) = \boxed{}$

(B) Garden borders

Write an expression for the perimeter of each garden, then simplify your expression.

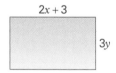

a $2x + 3$ $3y$

$p = \mathbf{2x + 3 + 3y + 2x + 3 + 3y}$

$= \underline{}$

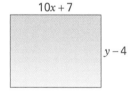

b $10x + 7$ $y - 4$

$p = \underline{}$

$= \underline{}$

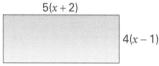

c $5(x + 2)$ $4(x - 1)$

$p = \underline{}$

$= \underline{}$

(C) Pyramids

The expression in each box is found by adding the expressions in the two boxes below it. Find the missing expressions in these. Simplify them.

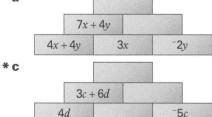

a

$7x + 4y$

$4x + 4y$ $3x$ ^-2y

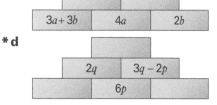

b

$3a + 3b$ $4a$ $2b$

*c

$3c + 6d$

$4d$ ^-5c

*d

$2q$ $3q - 2p$

$6p$

36 Substituting into Expressions

Let's look at ...
● evaluating expressions by substituting values

These are the points you need to know.

✓ We **evaluate** an expression by **substituting** values for the unknown. We follow the rules for **order of operations**.

   Example If $x = {}^-1$ and $y = 3$, then

   $$20 - x = 20 - {}^-1 \qquad\qquad 5y^3 = 5 \times 3^3$$
   $$= 20 + 1 \qquad\qquad\qquad = 5 \times 27$$
   $$= 21 \qquad\qquad\qquad\quad = 135$$

HELP

Need help with order of operations? Go to page 7

(A) Hexamazes

If $x = 3$ find the value of each of the following.
Shade the answers in the hexamaze to find the path from start to finish.

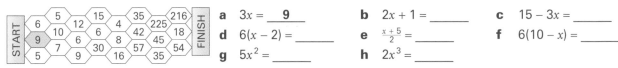

a $3x = \underline{\quad 9 \quad}$

b $2x + 1 = \underline{\qquad}$

c $15 - 3x = \underline{\qquad}$

d $6(x - 2) = \underline{\qquad}$

e $\frac{x + 5}{2} = \underline{\qquad}$

f $6(10 - x) = \underline{\qquad}$

g $5x^2 = \underline{\qquad}$

h $2x^3 = \underline{\qquad}$

If $p = 2$, $q = 4$, $r = 5$ and $s = 9$ evaluate the following.
Shade the answers in the hexamaze.

i $p + r - q = \underline{\qquad}$

j $s^2 - r = \underline{\qquad}$

k $p^2 + q^2 = \underline{\qquad}$

l $3r^2 - 1 = \underline{\qquad}$

m $2q^2 - p = \underline{\qquad}$

n $r - q - p = \underline{\qquad}$

o $q^2 - p^2 = \underline{\qquad}$

p $\sqrt{r^2 - q^2} = \underline{\qquad}$

q $\sqrt[3]{\frac{q^2}{2}} = \underline{\qquad}$

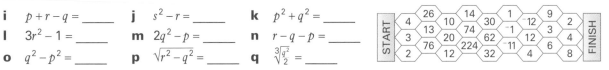

Evaluate the following and shade the answers in the hexamaze.

If $y = 3 \cdot 5$ find **r** $2y + 1 = \underline{\quad}$ **s** $4(7 - y) = \underline{\quad}$ **t** $15 - 3y = \underline{\quad}$

If $w = 4 \cdot 2$ find **u** $5w = \underline{\quad}$ **v** $10w - 6 = \underline{\quad}$ **w** $7(w - 4) = \underline{\quad}$

If $z = {}^-6$ find **x** $3z + 20 = \underline{\quad}$ **y** $\frac{z + 18}{z} = \underline{\quad}$ **z** $\frac{z - 2}{z + 4} = \underline{\quad}$

(B) Tiling

The number of tiles in shape n is given by $n^2 - 1$.

Shape 1   **Shape 2**   **Shape 3**   **Shape 4**

Find the number of tiles in shape number

a 6 ____   **b** 12 ____   **c** 20 ____   ***d** 50 ____

(C) Magic or not?

In a magic square every row, column and diagonal must add to the same total.

$x-y$	$x+y-z$	$x+z$
$x+y+z$	x	$x-y-z$
$x-z$	$x-y+z$	$x+y$

If $x = 5$, $y = 3$ and $z = 2$ use the expressions on the left to complete Square A.

Is Square A magic? _____

How do you know? _____

_____

Square A

How did you find this?   **EASY**   **OK**   **HARD**

37 Substituting into Formulae

Let's look at ...
● substituting known values into a formula
● solving an equation to find the unknown

These are the points you need to know.

✓ When we substitute values for unknowns into a **formula** we sometimes need to solve an equation.

Example   The formula for finding the speed, S, in km/h, is $S = \frac{D}{T}$

D is the distance, in km, and T is the time taken, in hours.
If $S = 80$ km/h and $T = 2$ hours, then
$$80 = \frac{D}{2}$$
$$80 \times 2 = D$$
$$\boldsymbol{D = 160 \text{ km}}$$

A Getting started

a   The number of litres of paint, ℓ, required to cover a room is $\ell = \frac{A}{16}$.
A is the area of the room in square metres.
Find ℓ if   **i** $A = 32$, _$\ell =$___   **ii** $A = 40$, _$\ell =$___   **iii** $A = 76$ _$\ell =$___

b   A mother feeds her baby milk according to this formula. $C = \frac{180\,w}{n}$.
C is the volume of milk in mℓs, w is the baby's mass in kg and n is the number of feeds the baby has each day.
How much milk should the mother give her baby if
i $w = 7$, $n = 7$ _____   **ii** $w = 9$, $n = 6$ _____   **iii** $w = 8\cdot2$, $n = 4$ _____

c   The area of this trapezium is given by
$$A = \frac{h(a + b)}{2}$$

Find the area if
i $h = 3$ cm, $a = 4$ cm, $b = 6$ cm   _$A =$___   **ii** $h = 3\cdot4$ m, $a = 5\cdot2$ m, $b = 7$ m   _$A =$___

d   The pay rate, P pounds, of a delivery girl is given by $P = 3n + 0\cdot5d$.
n is the number of deliveries, d is the distance travelled in km.
What is the pay rate for   **i** $n = 4$, $d = 31$ _$P =$___   **ii** $n = 7$, $d = 95\cdot6$ _$P =$___

B A bit trickier

a $A = bh$ gives the formula for the area of a parallelogram.
Find   **i** b if $A = 36$ cm^2 and $h = 3$ cm   _$b =$___
       ii h if $A = 175$ mm^2 and $b = 25$ mm   _$h =$___

b   Robyn calculates her earnings, £P, each Saturday using the formula $P = 9b - 20$.
b is the number of hand-made bears she sells at her local market.
How many bears did she sell if her earnings were.
i £7 _____   **ii** £106 _____   **iii** £241 _____   **iv** £304? _____

c   The formula for the sum of the angles in a polygon, $T°$, is
$T = 180(n - 2)$ where n is the number of sides.

How many sides has a polygon if
i $T = 360°$ _____   **ii** $T = 900°$ _____   **iii** $T = 4320°$? _____

*d   The volume of this cube is given by the formula $V = \ell^3$.

Find ℓ if   **i** $V = 8$ m^3 $\ell =$ _____   **ii** $V = 216$ m^3 $\ell =$ _____   **iii** $V = 2197$ m^3 $\ell =$ _____

38 Writing Expressions

Let's look at ...
● using information given to write expressions

✓ We can use information to **write expressions**.

Example   April has *n* pairs of shorts and some tops.
She has three times as many tops as shorts.
An expression for the number of tops she
has is $3 \times n = 3n$.
April threw out half of her pairs of shorts
and gave one pair to a cousin.
An expression for the number of pairs of
shorts left is $n - \frac{n}{2} - 1$.

These are the points
you need to know.

| Number of pairs originally | Number of pairs thrown out | Pair given to cousin |

This expression could have
been written as $n - (\frac{n}{2} + 1)$.

(A) *Work it out*

Write all expressions in
their simplest form.

1   Mark works for *x* hours each day.

   a   Write an expression for the number of hours he works in 15 days. _____

   b   One week he works for 5 days but on the fifth day he leaves work 3 hours early to go on holiday.
   Write an expression for the number of hours worked this week. _____

2   Odette has *p* pounds in her pocket.
She has three times as much money in her wallet.

   a   How much money does she have in her wallet? _____

   b   How much money does she have in her pocket and her wallet combined? _____

   c   Odette buys a £3 sandwich. How much money does she have left? _____

3   It takes Riordan *t* minutes to walk to school.
It takes him half the time if he travels by bus.

   a   Write an expression for the time it takes Riordan to get to school on the bus. _____

   b   Yesterday the bus broke down. It took 45 minutes to get it going again.
   How long did Riordan take to get to school on the bus yesterday? _____

4   Sheena's dog is *y* years old.

   a   Sheena's cat is 3 times as old as her dog. Write an expression for the cat's age. _____

   b   Sheena's sister is 4 years older than her cat. How old is her sister? _____

 *** c**   Sheena's uncle is twice as old as her sister. Write an expression for her uncle's age. _____

5   Joe has some bottles of lemonade.
Each bottle has a capacity of *n* mℓs.
Write an expression for the total capacity of each of these.

a **b** *** c** Joe drank one third of this bottle.

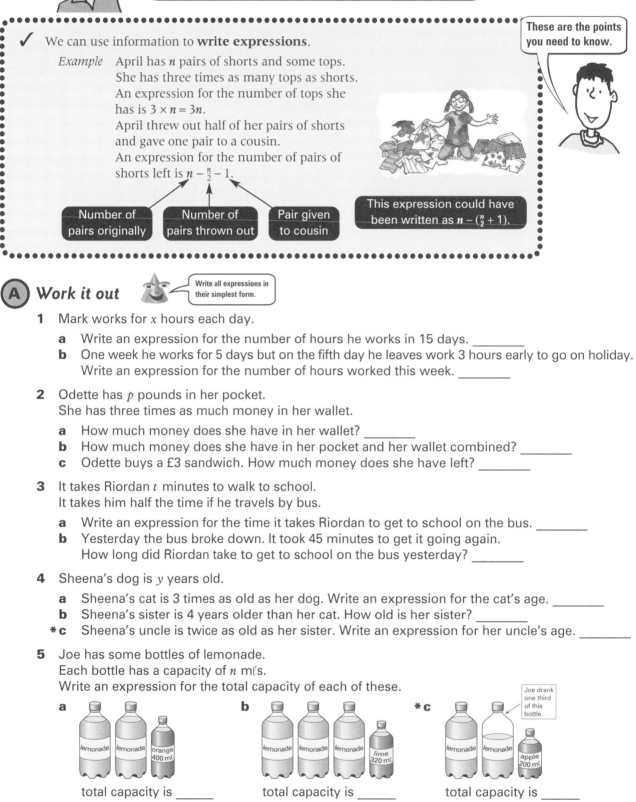

total capacity is _____   total capacity is _____   total capacity is _____

 *** d**   Joe gives his family $2\frac{1}{2}n + 100$ mℓs of drink.
Draw a picture to show what he could have given them.

How did you find this?   EASY   OK   HARD

39 Harder Writing Expressions

Let's look at ...
- using information given to write expressions
- showing that different expressions are equivalent

These are the points you need to know.

✓ We can often **write two equivalent expressions** from the same information.

Example A courtyard was built around a tree.

- Jesamine wrote this expression for the area of the courtyard.
$12 \times 8 - 3 \times x = 96 - 3x$

- Ronan divided the courtyard into two rectangles.
He wrote this expression for the area of the courtyard.
$(12 - x) \times 3 + 5 \times 12 = 3(12 - x) + 60$

We can show that these two expressions are equivalent by simplifying Ronan's expression.
$3(12 - x) + 60 = 36 - 3x + 60 = 96 - 3x$

A Flynn's farm

Flynn has horses, cows and sheep on his farm.
He has 3 times as many cows as horses.
He has 20 more sheep than cows.

a Assume that there are n horses.
  i Write an expression for the number of cows there are. _____
  ii Write an expression for the number of sheep there are. _____

b Assume that there are x cows.
  i Write an expression for the number of sheep. _____
  ii Write an expression for the number of horses. _____

c Assume that there are p sheep.
  i Write an expression for the number of cows. _____
  ii Write an expression for the number of horses. _____

B Red cross

a Write a simplified expression for the perimeter of this cross. _____

b A square has the same perimeter as this cross.
Write an expression for the length of the side of the square. _____

*c Write an expression for the area of the square. _____

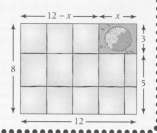

*C Kerry's kitchen

Kerry said that the area of her kitchen is $11y - 8$.

a Is Kerry's expression correct? _____
Explain why or why not. _____

b I divided Kerry's kitchen into these two rectangles.
Use this diagram to write a different expression for the area of Kerry's kitchen. _____

c Divide up Kerry's kitchen into two different rectangles.
Use this to write a third expression for the area of Kerry's kitchen. _____

d Simplify your expressions from **b** and **c** to show that they are equivalent to Kerry's expression $11y - 8$.
Simplify **b**. _____
Simplify **c**. _____

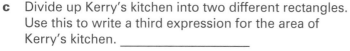

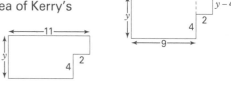

40 Writing Equations

Let's look at ...
● **using information given to write equations**

These are the points you need to know.

✓ We can **write equations** from information given.

Example Becks has *n* computer games and Jo has *m*.
If Becks gave away 3 games, she would have four times as many as Jo.
We can write an equation for this.
If Becks gives away 3 games, she will have $n - 3$ left.
This is the same as four times as many as Jo. Four times as many as Jo is $4m$.
So $n - 3 = 4m$

> Remember an equation always has an equals sign.

A *At the market*

1 Lucy bought *f* pieces of fruit and *v* vegetables at her local market.
Fill in the gaps.
   a Altogether she bought 48 pieces of fruit and vegetables. $f + v =$ _____
   b If Lucy bought 4 more pieces of fruit, she would have 16 pieces of fruit.
   $f +$ _____ $= 16$
   c Lucy bought 3 times as many vegetables as fruit. $v =$ _____

2 Dominic bought a hat for £*h* and a shirt for £*t*.
Fill in the gaps.
   a The hat and the shirt cost £80 altogether. _____ + _____ = £80
   b Dominic could have bought 10 shirts for £450. _____ = £450
   c If the hat cost £10 more it would have been the same price as the shirt. $h +$ _____ = _____

B *It's in the post*

In a month, the Sanders family received *b* bills and *l* letters.
Write an equation for each of these.

   a The Sanders received 5 more bills than letters. _____

   b Half the number of bills is the same as two thirds of the number of letters. $\frac{1}{2}b =$ _____

   c If the Sanders had received 10 less letters, this would have been the same as one quarter of the number of bills. _____

C *Stormy night*

During a storm *p* homes lost power and *t* homes had trees blown over.
Write an equation for each of these.

1 **a** Four times as many homes had trees blown over than lost power. _____
   b If 500 more homes had trees blown over this would be six times the number of homes which lost power. _____
   c One-fifth of the number of homes which lost power is the same as one-twentieth of the number of houses with trees blown over. _____

*2 During the storm *f* fire engines, *p* police cars and *a* ambulances were called out.
Write an expression for each of these.
   a The total number of fire engines and ambulances was the same as the number of police cars. _____
   b If there were four less ambulances, this would be the same as one-fifth of the number of fire engines. _____
   c One-third of the number of fire engines added to the number of ambulances is the same as half of the number of police cars. _____

How did you find this? EASY OK HARD

41 Writing and Finding Formulae

> **Let's look at ...**
> ● using information given to write formulae

✓ We can **write formulae** using given information.

Example A roast chicken takes 40 minutes per kilogram to cook.
2 kilograms would take 40×2 minutes to cook.
3 kilograms would take 40×3 minutes to cook.
n kilograms would take $40 \times n$ minutes to cook.

We could write a formula for the time taken, T, in minutes, to cook n kilograms. **$T = 40n$**

time (min)   number of kilograms

A *After school*

Write a formula for each of these

a The number of metres travelled, m, is found by multiplying the number of kilometres travelled, n, by 1000. $m =$ _____

b The number of spoons of drinking chocolate required, s, is found by multiplying the number of cups, c, by 2. $s =$ _____

c The time spent watching television, t is found by subtracting the time spent doing homework, h, from 4. $t =$ _____

d The mass of a school bag, w, is found by multiplying the number of books, b, by 400 and then adding 900. $w =$ _____

B *Sale yards*

1 pen   2 pens   3 pens

Sheep pens are made using sections of fence.
Paul wants to know if there is a formula for finding the number of fence sections, f, needed to make p pens.

a Draw diagrams of 4 and 5 pens.

b Complete this table.

*c Find the formula. $f =$ _____

Number of pens, p	1	2	3	4	5	...
Number of fence sections, f	3	5				...

C *Kelsey's hexagons*

Kelsey makes pot stands from hexagonal tiles.
She wants to know if there is a formula for calculating the number of white tiles, w, that are needed for a pot stand with p purple tiles.

a Draw a diagram of a pot stand with 4 purple tiles.

b Complete this table.

*c Find the formula. $w =$ _____

Number of purple tiles, p	1	2	3	4	5	...
Number of white tiles, w	6					...

42 Writing and Solving Equations

Let's look at ...
● solving equations using inverse operations

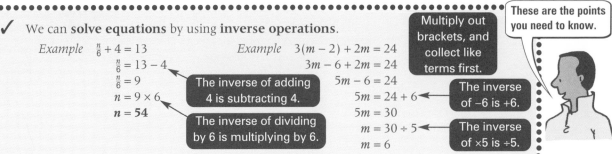

✓ We can **solve equations** by using **inverse operations**.

Example $\frac{n}{6} + 4 = 13$

$\frac{n}{6} = 13 - 4$

$\frac{n}{6} = 9$

$n = 9 \times 6$

$n = 54$

The inverse of adding 4 is subtracting 4.

The inverse of dividing by 6 is multiplying by 6.

Example $3(m - 2) + 2m = 24$

$3m - 6 + 2m = 24$

$5m - 6 = 24$

$5m = 24 + 6$

$5m = 30$

$m = 30 \div 5$

$m = 6$

Multiply out brackets, and collect like terms first.

The inverse of -6 is $+6$.

The inverse of $\times 5$ is $\div 5$.

These are the points you need to know.

(A) Lots of unknowns

Solve these equations using inverse operations. Show your working.

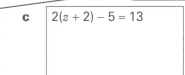

Start by expanding the brackets.

a $3x - 1 = 11$

b $5(y + 3) = 45$

c $2(z + 2) - 5 = 13$

Solve these equations. You will need a separate sheet of paper for your working.

d $8n + 1 = 49$ $n = \_\_\_$

e $102 = 5a + 7$ $\_\_\_$

f $7(d + 4) = 49$ $\_\_\_$

g $6(w - 4) = 42$ $\_\_\_$

h $41 = 3(y + 5) + 2$ $\_\_\_$

i $4(5c - 8) = 32$ $\_\_\_$

j $3(n + 9) - 4 = 29$ $\_\_\_$

k $5p + 6 - 2p + 3 = 24$ $\_\_\_$

*l $14\cdot8 = 9\cdot8m - 14\cdot6$ $\_\_\_$

*m $1\cdot6p + 7\cdot6 = 22$ $\_\_\_$

*n $2(x - 6) + 3x = 28$ $\_\_\_$

*o $10(t + 2) - 5(t - 1) = 37$ $\_\_\_$

(B) Write your own

Show your working for these questions.

Write and solve equations for each of these.

a Muffins usually cost £m. In a sale they cost £1 less than usual.
In this sale 8 muffins cost £16. How much do muffins usually cost?

You will need to use brackets when you write your equation.

Write an expression for the cost of 1 muffin in the sale. _____

Multiply this by 8. _____ × (_____   ← Cost of 1 muffin

Now write an equation for the cost of these 8 muffins. _____ = _____

Solve this equation to find the usual cost of a muffin. _____

b Robert has x CDs. Stephen has 9 more CDs than Robert. Terry has 7 times as many CDs as Robert. Altogether they have 45 CDs. How many does each person have?

(C) Pyramid

In these pyramids each number is the sum of the two numbers immediately above it.
Write an equation for each pyramid, then use this equation to find n.

a

| 10 | n | 20 |
| 60 | | |

$n = _____$

b

| 17 | n | 21 |
| 84 | | |

$n = _____$

c

| 29 | n | 18 |
| 85 | | |

$n = _____$

How did you find this?   EASY   OK   HARD

43 Solving Equations by Transforming Both Sides

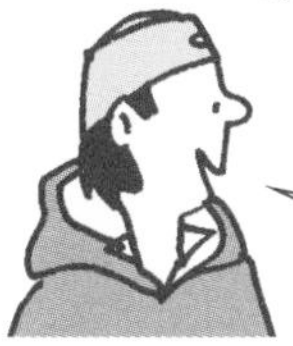

Let's look at ...
- an alternative method for solving equations

✓ When unknowns are on both sides of the equation we **transform both sides** to solve the equation.

Example

$5x + 1 = 3x + 3$

$5x + 1 - \mathbf{3x} = 3x + 3 - \mathbf{3x}$ **subtract 3*x* from both sides**

$2x + 1 = 3$

$2x + 1 - \mathbf{1} = 3 - \mathbf{1}$ **subtract 1 from both sides**

$2x = 2$

$\frac{2x}{2} = \frac{2}{2}$ **divide both sides by 2**

$\mathbf{x = 1}$

Subtract the smaller amount of unknown from both sides

Always do the *same thing* to both sides.

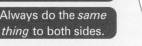

These are the points you need to know.

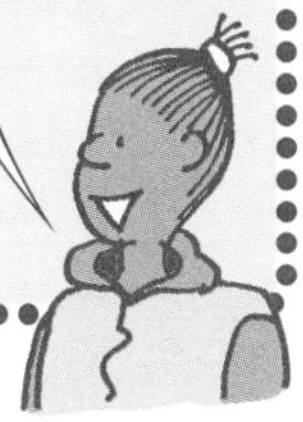

A Show your skills

Solve these equations by transforming both sides. Show your working.

a $\frac{x}{2} + 3 = 9$

b $4w - 2 = 3w + 2$

c $5n + 3 = 19 - 3n$

B Hold your breath!

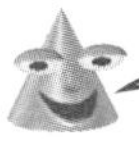

You will need a separate sheet of paper for your working.

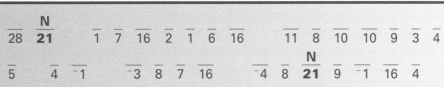

_ _ N _ _ _ _ _ _ _ _ _ _ _ _ _ _
28 **21** 1 7 16 2 1 6 16 11 8 10 10 9 3 4

_ _ _ _ _ _ _ _ _ N _ _ _ _
5 4 ⁻1 ⁻3 8 7 16 ⁻4 8 **21** 9 ⁻1 16 4

Solve these equations. Write the letter beside each equation above its answer in the box.

N $x + 7 = 28$, $x = \mathbf{21}$	**V** $8x = 56$	**G** $3x - 2 = 16$	**U** $25 = 7 + 2x$
S $\frac{3x}{2} = 6$	**I** $\frac{5x}{2} = 20$	**E** $3 + \frac{x}{2} = 11$	**O** $\frac{x}{4} - 1 = 6$
P $4x + 3 + 2x = 21$	**L** $9x - 5 - 2x = 30$	**R** $6x = 5x + 2$	**A** $7x - 6 = 5x - 4$
C $4x - 8 = 2x + 12$	**H** $8 + 2x = 41 - x$	**T** $5x + 4 = 3x + 2$	**F** $2 - x = 2x + 11$
M $10 + 2x = 18 + 4x$			

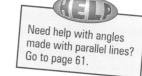

HELP
Need help with angles made with parallel lines? Go to page 61.

C Geometric gems

Write and solve an equation to find the value of n. Show your working.

a

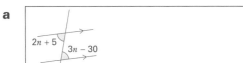

2n + 5
3n − 30

b

5n + 2
3n + 46

c

4n − 7
23 − n

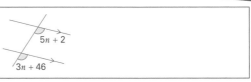

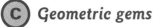

***d** These two triangles have the same perimeter.

n
n
4n − 6
2n − 1
2n − 1
n

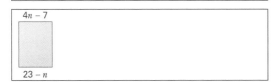

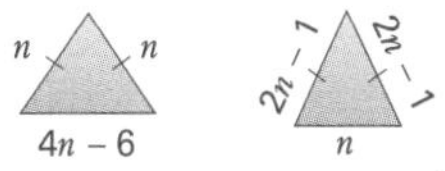

How did you find this? EASY OK HARD

44 Equations and Graphs

Let's look at ...
● solving equations using a graph

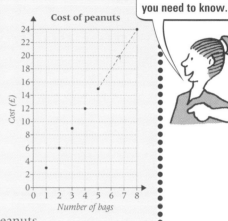

Cost of peanuts

These are the points you need to know.

✓ One bag of peanuts costs £3.
We can write a table to show how much 2, 3, 4, 5 ... bags cost.

Number of bags	1	2	3	4	5	...
Cost (£)	£3	£6	£9	£12	£15	...

✓ We can work out the ratio   *Cost : Number bought*
for each pair of values on the table.
All of the ratios simplify to *3 : 1*, so we
know that cost is directly proportional
to the number of bags bought.

£6 : 2 = 3 : 1,
£9 : 3 = 3 : 1

✓ The graph shows Cost verses Number of bags bought.
We can extend the points plotted to find the cost of 8 bags of peanuts.
8 bags of peanuts cost £24.

A Weed and mow

Rebecca earns £5 for 1 hour's work.

a Complete this table.

Number of hours worked	1	2	3	4	5	...
Money earned (£)	5					

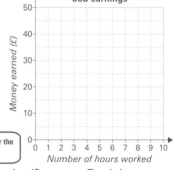

Job earnings

b Work out the ratio   *Hours worked : Money earned*   for each
pair of values on the table. What do you
notice? _____

Hint: For the second pair the ratio is 2 : 10 = 1 : 5.

c Is the money earned directly proportional to the number of hours worked? _____ Explain _____
_____

d Plot the graph of *money earned* versus *hours worked*.
Do the points lie on a straight line? _____

*e Write a formula for the relationship between *money earned* (y) and *hours worked* (x). $y =$ _____

f How much would Rebecca earn if she worked for 9 hours? _____ Explain how you found this.
_____

B Mix it up

To make orange paint Katherine mixes 5 spoons of yellow paint with 3 spoons of red paint.

a Complete this table.

Yellow paint	5	10	15	20	25	...
Red paint	3					...

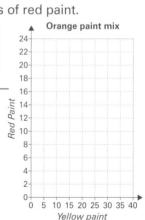

Orange paint mix

b Is the ratio   *Yellow paint : Red paint*   constant? _____ What is it? _____

c Plot the graph of *Yellow paint* versus *Red paint*.

*d Which of the following is the formula for the relationship between
red paint (y) and *yellow paint* (x)? _____
A $y = 5x$ **B** $y = 3x$ **C** $y = \frac{3}{5}x$ **D** $y = \frac{5}{3}x$

*e How much red paint should Katherine mix with
i 35 spoons of yellow paint _____
ii 150 spoons of yellow paint? _____

How did you find this?   EASY   OK   HARD

45 Generating Sequences

Let's look at ...
● **three different ways to generate a sequence**

These are the points you need to know.

✓ We can write a sequence from a **flow chart**.

✓ We can generate a sequence by **multiplying or dividing by a constant number**.

Examples   1, 2, 4, 8, 16, 32, ...   multiplying by 2

729, ⁻243, 81, ⁻27, 9, ⁻3, ...   dividing by ⁻3

✓ We can generate a sequence by **counting forwards or backwards in increasing or decreasing steps**.

Examples   1, 2, 4, 7, 11, ...   counting on in steps of 1, 2, 3, 4, ...
               +1 +2 +3 +4

A Follow the flow

Write down the sequence given by this flow chart. _____

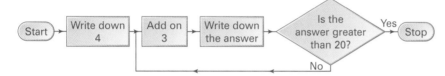

Start → Write down 4 → Add on 3 → Write down the answer → Is the answer greater than 20? — Yes → Stop / No

B First five and next two

Write down the first five terms of these sequences.

a **1st term** 1   **rule** multiply by 3 _____

b **1st term** 625   **rule** divide by 5 _____

c **1st term** 4   **rule** multiply by ⁻2 _____

d **1st term** 16   **rule** divide by 4 _____

Predict the next two terms in these sequences.

e 5, 10, 20, 40, _____, _____   **f** 96, 48, 24, 12, _____, _____   **g** 1, ⁻3, 9, ⁻27, _____, _____

C First six

Marie started at 2 and counted on in steps of 3, 5, 7, 9, ... She got this sequence
Write down the first six terms of these sequences.

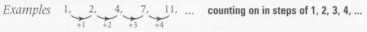
2, 5, 10, 17, 26, ...
  +3 +5 +7 +9

a Start at 4 and count forwards by 1, 2, 3, 4, ... _____

b Start at 20 and count forwards by 2, 4, 6, 8, ... _____

c Start at 50 and count backwards by 1, 3, 5, 7, ... _____

D Leisure time

a John started a new book. The first day he read 1 page. Each day he read twice as many pages as the day before.
Write down how many pages he read on each of the first six days. _____

b Robert went on a fishing holiday. Each day he increased the number of fish he caught by 1, 2, 3, 4, 5, ... On the first day he caught 2 fish.
i   Write down how many fish he caught on the first six days. _____
ii On which day did he catch 23 fish? ____

How did you find this?   EASY   OK   HARD

46 Continuing Sequences

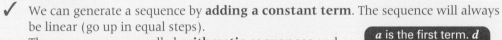

Let's look at ...
- writing arithmetic sequences
- predicting the next few terms of a sequence

These are the points you need to know.

✔ We can generate a sequence by **adding a constant term**. The sequence will always be linear (go up in equal steps).
These sequences are called **arithmetic sequences** and are generated by starting with a number *a*, and adding a constant number, *d*.

a is the first term. *d* is the amount added to get the next term.

Example If *a* = 4 and *d* = ⁻2 the sequence is 4, 2, 0, ⁻2, ⁻4, ...
−2 −2 −2 −2

✔ We can **predict** the next few terms of a sequence by looking for a **pattern**.
To predict with certainty, we must know the **rule for the sequence**.

Example 1, 2, 3 could continue as 1, 2, 3, 4, 5, 6, ... (adding 1) or as
1, 2, 3, 5, 8, 13, ... (adding the two previous terms).

Ⓐ Match it up

Draw a line to match each arithmetic sequence with its starting number *a*, and constant *d*.

a 2, 7, 12, 17, ... • • *a* = 4, *d* = 2   **e** 5, 9, 13, 17, ... • • *a* = 5, *d* = 6
b 4, 6, 8, 10, ... • • *a* = ⁻2, *d* = 4   **f** 5, ⁻1, ⁻7, ⁻13, ... • • *a* = 5, *d* = ⁻6
c ⁻2, 2, 6, 10, 14, ... • • *a* = 2, *d* = 5   **g** 5, 1, ⁻3, ⁻7, ... • • *a* = 5, *d* = 4
d 4, 2, 0, ⁻2, ⁻4, ... • • *a* = 4, *d* = ⁻2   **h** 5, 11, 17, 23, ... • • *a* = 5, *d* = ⁻4

Ⓑ First six

Write down the first six terms of these arithmetic sequences.

a *a* = 50, *d* = 5   **50, 55,** _____   **b** *a* = 0·3, *d* = 0·1 _____
c *a* = ⁻10, *d* = 3 _____   **d** *a* = 3, *d* = ⁻1 _____

Ⓒ Missing links

Fill in this diagram so that the four sequences shown by the arrows are arithmetic.

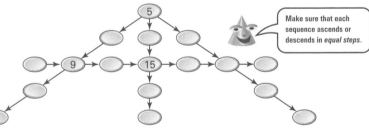

Make sure that each sequence ascends or descends in *equal steps*.

Ⓓ Puzzle time

a The fourth term of a sequence is 3. If *d* is 2, what is *a*? _____
∗b A sequence is arithmetic. *a* = 5. One of the terms is 25.
Find all of the values *d* could have. _____

Ⓔ What comes next?

Predict the next three terms of these sequences. Say if the sequence continues in equal or unequal steps.

a 7, 14, 21, 28, ___, ___, ___   _equal_   **b** 3, 6, 12, 24, ___, ___, ___ _____
c 202, 198, 194, 190, ___, ___, ___ _____   **d** 3, 4, 6, 9, 13, ___, ___, ___ _____
e 256, 128, 64, 32, 16, ___, ___, ___ _____   **∗f** 2, 2, 4, 6, 10, 16, ___, ___, ___ _____

Continue this sequence in two different ways. Write your rule for each.

∗g 2, 3, 5, ___, ___, ___   _____
2, 3, 5, ___, ___, ___   _____

How did you find this?   EASY   OK   HARD

47 Writing Sequences from Rules

Let's look at ...
- **using term-to-term rules to write sequences**
- **using $T(n)$, the rule for the nth term, to write sequences**

✓ We can write a sequence if we know the **term-to-term rule** or the **rule for the nth term**. $T(n)$ is the **notation** for the nth term (general term).

Examples **1st terms** 2, 3   **term-to-term rule** add the two previous terms

generates the sequence   2, 3, 5, 8, 13, ...

$T(n) = 50 - 5n$ generates   45, 40, 35, 30, 25, ...

These are the points you need to know.

when $n = 1$, $T(n) = 50 - 5 \times 1 = 45$
$n = 2$, $T(n) = 50 - 5 \times 2 = 40$
$n = 3$, $T(n) = 50 - 5 \times 3 = 35$ and so on.

(A) Missing numbers

Use the given term-to-term rules to find the missing numbers.

Term-to-term rule

a add 6   4, □, □, □, □, □, □

b multiply by 10   5, □, □, □, □, □

c divide by 2   □, □, 24, □, □, □, □

d add consecutive odd numbers starting with 1   7, □, □, □, □, □

e add previous two terms   1, 4, □, □, □, □, □

*f add consecutive even numbers starting with 2   □, □, □, 18, □, □

(B) Table of terms

Complete this table.

$T(n)$	$T(1)$	$T(2)$	$T(3)$	$T(4)$	$T(5)$	...	$T(20)$
$2n + 1$	$\overset{2 \times 1 + 1}{3}$	$\overset{2 \times 2 + 1}{5}$				...	
$15 - n$						...	
$35 - 3n$						...	
$\frac{1}{2}n + 3$						...	

(C) Match it up

$T(n) = 3n + 1$ and 'first term 4, **rule** add 3' both give the sequence **4, 7, 10, 13, 16, ...**.
Match up these rules and give the first five terms for each.

$T(n) = 3n + 1$ •   • first term 2, rule add 1   _____

$T(n) = 1 + n$ •   • first term 1, rule add 2   _____

$T(n) = 2n - 1$ •   • first term 4, rule add 3   **4, 7, 10, 13, 16**

$T(n) = 20 - 2n$ •   • first term 1·5, rule add 2   _____

$T(n) = 2n - 0·5$ •   • first term 18, rule subtract 2   _____

*(D) Puzzle it

a Jessica wrote down the first five numbers in a sequence.
She said, '*The rule for my sequence is add □, where □ is greater than 2.*
All the numbers in my sequence are odd.
The first five numbers are all between 0 and 20.'
What might Jessica's rule and her first term be? _____

b Jack wrote down the first five numbers of another sequence.
He said, '*The rule for my sequence is subtract □, where □ is greater than 0.*
All the numbers in my sequence end in 4.
The first five numbers are all between 0 and 50.'
What is Jack's rule and his first term? _____

48 Describing Linear Sequences

Let's look at ...
● describing the rule for a linear sequence

These are the points you need to know.

✓ We can **describe a sequence** by looking at a rule for the *n*th term.
A linear sequence has the rule $T(n) = an + b$ where *a* and *b* are any numbers.

Example $T(n) = 3n + 1$ generates numbers with a difference of 3, starting at 4. All terms are one more than a multiple of 3.

The number, *a*, gives the difference between consecutive terms.

Ⓐ Match them up

$T(n) = 2n + 1$ has a difference of 2 between terms and the terms are all one more than a multiple of 2.
Draw a line to match each rule with a description.

Try to find the answer just by looking at the expression.

$T(n) = 3n$ ●

$T(n) = 3n + 1$ ●

$T(n) = 3n + 3$ ●

$T(n) = 3n - 6$ ●

$T(n) = 7 - 3n$ ●

● multiples of 3, starting at ⁻3

● multiples of 3, starting at 6

● descending numbers starting at 4, with a difference of 3 between terms

● multiples of 3, starting at 3

● ascending numbers starting at 4, with a difference of 3 between consecutive terms, all one more than a multiple of 3

Ⓑ Sort them out

a Write each of these rules in the correct box below.

$T(n) = 2n + 4$ $T(n) = 5n - 3$ $T(n) = 7n + 4$ $T(n) = n + 12$
$T(n) = 6n - 1$ $T(n) = 7 - 2n$ $T(n) = 8 - 3n$ $T(n) = 10 - n$

Difference between terms is odd	Difference between terms is even

b Now write each of these rules in the correct box.

$T(n) = 6n$ $T(n) = 6n + 3$ $T(n) = 6n + 6$ $T(n) = 18 - 6n$

Terms are multiples of 6	Terms are not multiples of 6

Ⓒ Describe them

Describe the sequence generated by these. Say whether it is ascending or descending.
Give the difference between terms, the starting number and any relevant information about multiples.

a $T(n) = 4n - 1$ _____

b $T(n) = 66 - 6n$ _____

*Ⓓ Write the rule

a Complete this rule in three different ways, so that the terms will be multiples of 7.

   $T(n) = 7n + \boxed{}$ $T(n) = 7n + \boxed{}$ $T(n) = 7n + \boxed{}$

Complete these rules so that these sequences will be generated.

b Ascending even numbers, starting at 2 $T(n) = \boxed{}$

c Ascending odd numbers, starting at 5 $T(n) = \boxed{}$

d 10 times table backwards, starting at 100 $T(n) = \boxed{}$

How did you find this?   EASY   OK   HARD

49 Sequences in Practical Situations

Let's look at ...
● finding the rule for the *n*th term in a practical situation

These are the points you need to know.

✓ We can find the rule for the *n*th term in a **practical situation**.

Example

shape 1   shape 2   shape 3

● The sequence generated by the number of squares in shapes 1, 2, 3, 4, ... is 4, 8, 12, 16, ...
● Each time a new shape is drawn 4 new squares are added.
  The expression for the number of squares in the *n*th shape is 4 times the shape number.
  We write this as 4*n*.

A Elegant Eggs

'Elegant Eggs' are trialling new cartons in which to package their eggs.

carton 1   carton 2   carton 3   carton 4

a Draw carton 4.

b What sequence is generated by the number of eggs in cartons 1, 2, 3, 4, ...? _____
Explain how you work out the next term of the sequence. _____

c Explain how you could find the number of eggs in carton *n*. Write an expression for this. _____
_____

B Happy Birthday

Janet decorates cakes with liquorice strips and sweets.

size 1   size 2   size 3

━ = liquorice strip
○ = sweet

Cake size	1	2	3	4	5
Number of liquorice strips	4				
Number of sweets	5				
Total number of decorations	9				

a Study the cake designs and complete the table.

b Explain how you would find the number of liquorice strips on a size *n* cake. _____
_____ Write an expression for this. _____

c Explain how you would find the number of sweets on a size *n* cake. _____
_____ Write an expression for this. _____

d Write an expression for the total number of decorations on a size *n* cake. _____

*C Cube it

John has a box of small wooden cubes. He paints a star on every face of every small cube, then glues them together as shown.

shape 1   shape 2   shape 3

a Complete this table.

Shape number	1	2	3
Number of stars on outside	6		

John picks up each shape and looks at every side when counting the stars.

b Explain how you would work out the number of stars on the outside of shape *n*. _____
_____

c Write an expression for the number of stars on the outside of shape *n*. _____

50 Finding the Rule for the *n*th Term

Let's look at ...
● writing an expression for the *n*th term of a sequence

✓ We can find a **rule for the *n*th term** by drawing a difference table.

Example For the sequence 5, 8, 11, 14, 17, ...

Term number	1	2	3	4	5
$T(n)$	5	8	11	14	17
Difference		3	3	3	3

The difference tells us the number that multiplies *n*.

The difference is **3** so the *n*th term is $3 \times n + ?$.
The first term is 5. This is $3 \times 1 + 2$. So $? = 2$
The rule for the *n*th term is $3n + 2$.

Check that the other terms are $3 \times n + 2$.

These are the points you need to know.

Ⓐ Who is right?

For each sequence, decide who has written the right rule.

	Sequence	Ann's Rule	Bob's Rule	Cathy's Rule	Who is right?
a	2, 4, 6, 8, 10, ...	$T(n) = n + 2$	$T(n) = 2n$	$T(n) = n - 2$	
b	5, 9, 13, 17, 21, ...	$T(n) = 4n + 5$	$T(n) = 5n + 4$	$T(n) = 4n + 1$	
c	9, 12, 15, 18, 21, ...	$T(n) = 3n + 6$	$T(n) = 3n + 9$	$T(n) = 9n + 3$	

Ⓑ What's the rule?

Find the rule for the *n*th term by completing the difference table.

a 4, 8, 12, 16, 20, ...

Term number	1	2	3	4	5
$T(n)$	4	8	12	16	20
Difference		4			

*n*th term = _____

b 3, 7, 11, 15, 19, ...

Term number	1	2	3	4	5
$T(n)$	3	7			
Difference					

*n*th term = _____

c 2, 8, 14, 20, 26, ...

Term number	1	2	3	4	5
$T(n)$					
Difference					

*n*th term = _____

d 0·2, 0·4, 0·6, 0·8, 1·0, ...

Term number	1	2	3	4	5

*n*th term = _____

* e 50, 45, 40, 35, 30, ...

Term number	1	2	3	4	5

*n*th term = _____

* f 2, ⁻4, ⁻10, ⁻16, ⁻22, ...

Term number		

*n*th term = _____

Ⓒ The hundredth

Find the 100th term of these sequences.

a 7, 13, 19, 25, 31, ... _____
b 30, 40, 50, 60, 70, ... _____
c 10, 18, 26, 34, 42, ... _____
d Explain how you found the 100th term in **c**. _____

_____

* Ⓓ Treasure trove

In an adventure movie doors are labelled in a sequence as shown above.

The hero is told that the treasure will be found behind the door numbered 205.
Does this door exist? _____ Explain how you know. _____

How did you find this? EASY OK HARD

51 Functions

Let's look at ...
● writing the inputs and outputs of a function in a table or as a mapping diagram

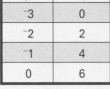
These are the points you need to know.

✓ A **function** can be written, using algebra, as an equation or a mapping.

Example 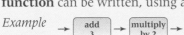 → add 3 → multiply by 2 → $y = 2(x + 3)$ **or** $x \rightarrow 2(x + 3)$
equation   mapping

✓ The inputs and outputs can be shown in a **table** or as a **mapping diagram**.

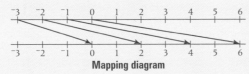
Mapping diagram

Input	Output
‾3	0
‾2	2
‾1	4
0	6

Table

✓ You should know these **properties of a mapping diagram**:

● Functions of the form $x \rightarrow x + c$ produce parallel lines.

Example $x \rightarrow x + 2$

● Mapping arrows for multiples, if extended backwards to meet the zero line, meet at a point.

Example $x \rightarrow 2x$

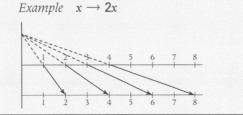

A Table it

Complete the input/output table for each function machine.

a $x \rightarrow$ multiply by 4 → add 1 → y

Input	Output
2	9
5	
10	
12	

b

33 →
63 → subtract 3 → divide by 10 → _
83 →
18 →

Input	Output
33	
63	
83	
18	

B Map it

Fill in these mapping diagrams for each function machine and input given.

a 0, 1, 2, 3, 4 → multiply by 2 → subtract 2 → y

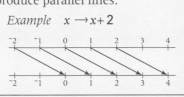

b ‾2, ‾1, 0, 1, 2 → add 1 → multiply by 2 → y

*c $y = \frac{x + 4}{2}$ for $x = 1, 0, ‾1, \frac{1}{2}, 2\frac{1}{2}$

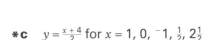

*C Sort it

Write each of these functions in the correct box below.

$x \rightarrow 3x$ $x \rightarrow x + 3$ $x \rightarrow x - 3$ $x \rightarrow 5x$ $x \rightarrow \frac{1}{3}x$

Lines on a mapping diagram will be parallel	Lines on a mapping diagram will meet on the zero line when extended backwards.

52 More Functions

Let's look at ...
- finding the rule, given the input and output
- properties of functions
- finding the input given the output

✓ If we are given the input and output we can **find the rule** for the function machine.

Example   2, 6, 4, 8 → [?] → [?] → 4, 16, 10, 22

Input (x)	2	4	6	8
Output (y)	4	10	16	22
Difference		6	6	6

Put the input and output in order

We use a difference table.
When the input increases by 2, the difference is 6.
When the input increases by 1, the difference will be 3.
The rule is $y = 3x - 2$.

These are the points you need to know.

✓ When **combining operations in a function** the operations follow the same rules as arithmetic operations.

✓ We can **find the input given the output** using an inverse function machine.

Example   ? → [divide by 2] → [add 4] → 28   48 ← [multiply by 2] ←24← [subtract 4] ← 28

The input is 48.

(A) Find the function

Fill in the functions for these.

Put the input and output in order first

a   2, 4, 3, 5, 1 → [ ] → 7, 9, 8, 10, 6

b   1, 3, 2, 5, 4 → [multiply by 3] → [ ] → 7, 13, 10, 19, 16

c   6, 4, 8, 2, 10 → [ ] → [ ] → 9, 5, 13, 1, 17

d   9, 5, 1, 7, 3 → [ ] → [ ] → 26, 14, 2, 20, 8

(B) Missing operations

a   Fill in this function machine in two different ways.

4, 9, 3, 6 → [multiply by 2] → [ ] → 10, 20, 8, 14   4, 9, 3, 6 → [ ] → [ ] → 10, 20, 8, 14

b   Fill in these missing operations.

16, 24, 8, 40 → [divide by 4] → [add 3] → [ ] → [ ] → 16, 24, 8, 40

Notice the input and output are the same.

(C) Yay or nay?

Are the two function machines equivalent or not? Circle the correct answer.

a
x → [add 2] → [add 6] → y
x → [add 8] → y

Equivalent
or
Not Equivalent

b
x → [multiply by 2] → [add 5] → y
x → [add 5] → [multiply by 2] → y

Equivalent
or
Not Equivalent

(D) Inverses and inputs

Complete the inverse function machines and then find the inputs for these.

a
__ , __ → [multiply by 3] → [subtract 4] → 17, 8
__ , __ ← [ ] ← [add 4] ← 17, 8

* What is the inverse function of
$x \rightarrow 3x - 4$? _____

b
__ , __ → [add 2] → [divide by 5] → 4, 7
__ , __ ← [ ] ← [ ] ← 4, 7

What is the inverse function of
$x \rightarrow \frac{x + 2}{5}$? _____

How did you find this?   [EASY]   [OK]   [HARD]

53 Graphing Linear Functions

Let's look at ...
● drawing a straight-line graph by plotting coordinate pairs

✓ **Remember** $y = 2x - 3$ $y = 20 - 3x$ are called **linear functions**.

A linear function has a y and an x term. It does not have x^2, x^3, ... terms.

To draw a **straight-line graph**
● find some coordinate pairs that satisfy the rule
● plot them on a grid
● draw a straight line through the points
● label the axes and line.

Each y coordinate is 2 times the x coordinate minus 1.

Example $y = 2x - 1$

x	0	2	⁻1
y	⁻1	3	⁻3

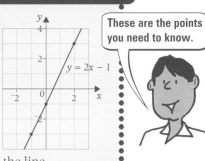

These are the points you need to know.

The coordinate pairs $(0, ⁻1)$, $(2, 3)$ and $(⁻1, ⁻3)$ are shown plotted.

The **equation of the line** in the example is $y = 2x - 1$.
The coordinate pairs of all points on the line satisfy the equation of the line.

(A) Get coordinated

Choose two coordinate pairs from the box that satisfy each function. The first one is done.

(2, 3)	(3, 8)	(4, 8)
~~(1, 4)~~	(5, 7)	(1, ⁻5)
(2, 7)	~~(2, 8)~~	(6, 7)
(2, ⁻12)	(5, 13)	(⁻2, ⁻7)

a $y = 4x$ **(1, 4)** , **(2, 8)**

b $y = x + 1$ _____ , _____

c $y = 2x + 3$ _____ , _____

d $y = 3x - 1$ _____ , _____

e $y = 12 - x$ _____ , _____

*f $y = 2 - 7x$ _____ , _____

(B) Graph it

a Complete these coordinate pairs for the function $y = 2x + 3$.

(2, ___), (1, ___), (0, ___), (⁻1, ___),

$\underbrace{}_{2 \times 2 + 3}$ $\underbrace{}_{2 \times 1 + 3}$

b Plot the coordinate pairs on the grid.
Draw a line through the points and label it.

c Does $y = 2x + 3$ go through the point
 i (2, 8) _____ **ii** (⁻2, ⁻1) _____ **iii** ($\frac{1}{2}$, 4)? _____

d Will the point (15, 35) lie on this line? _____
Explain how you know this _____
_____

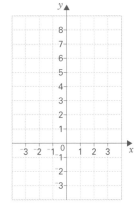

(C) Try this one

a Complete this table for the function $y = 4 - 2x$

x	0	1	2	3	⁻1
y					

b Plot the points given by the table on the grid.
Draw a line through the points and label it.

c Will the point (8, ⁻12) lie on this line? _____
Explain how you know this. _____
_____

*d $y = 4 - 2x$ goes through each of these points.
Complete the coordinate pairs.
(___, ⁻4), (⁻2, ___), (___, 1), (⁻0·5, ___), (___, ⁻3)

Hint: You could read some of these values off your graph.

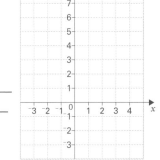

54 Graphing Linear Sequences

Let's look at ...
● plotting coordinate pairs from a linear sequence

✔ The graph of a **linear sequence** gives a set of points in an 'imagined' straight line.

Example

Term number	1	2	3	4
$T(n)$	⁻1	1	3	5

We plot the coordinate pairs (1, ⁻1), (2, 1), (3, 3), (4, 5).
These points lie on an 'imagined' straight line.
We do not draw a straight line through the points as there is no term number $1\frac{1}{2}$ or $2\frac{1}{4}$ for example.

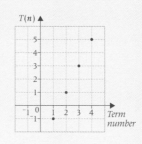

These are the points you need to know.

A Patrick's pattern

Patrick makes these shapes from toothpicks.

   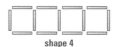

shape 1   shape 2   shape 3   shape 4

a Complete this table.

Shape number	1	2	3	4
Number of toothpicks	4			

b Write down the coordinate pairs from the table.

_____

c Plot these points on the grid.

d Do the points lie on an 'imagined' straight line? _____

e Should you draw a straight line through the points? _____
Why or why not? _____

_____

f Use your graph to find the number of toothpicks in shape 7. _____

g Will one of the shapes in Patrick's sequence need 20 toothpicks? _____
How do you know? _____

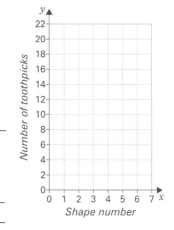

B Term by term

The nth term of a sequence is given by $T(n) = 6 - 2n$.

a Complete this table.

Term number	1	2	3	4
$T(n)$				

b Plot this sequence on the grid.

c Use your graph to find
$T(5) =$ _____ $T(6) =$ _____

d Is ⁻5 a term of this sequence? _____
How do you know this? _____

∗e What is the difference between the graph of the sequence
$T(n) = 6 - 2n$ and the graph of the function $y = 6 - 2x$? _____

_____

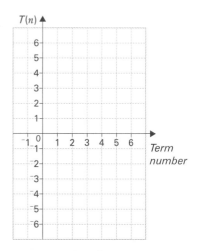

How did you find this? EASY   OK   HARD

55 Equations of Straight-line Graphs

Let's look at ...
- the equation for a straight line, $y = mx + c$
- interpreting $y = mx + c$

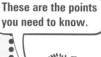

These are the points you need to know.

✓ $y = mx + c$ is the equation of a straight line.

✓ m represents the **gradient** or steepness of the line.
The greater the value of m, the steeper the slope.

If m is positive, the gradient is positive.
If m is negative, the gradient is negative.

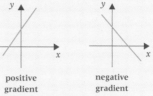

positive gradient   negative gradient

✓ c represents the **y-intercept** (where the graph cuts the y-axes).
The coordinates of the y-intercept are $(0, c)$ where c is given by the equation of the line.

Example $y = 2x - 1$ has a gradient of **2** and crosses the y-axis at $(0, {}^-1)$.

(A) *Eight of the best*

$y = 5x + 2$	$y = x + 3$	$y = 3 - 5x$	$y = 2 - 3x$	$y = 5x - 3$	$y = x - 2$	$y = 6x$	$y = 3$

Toby graphs each of these eight equations.

a   Which has the steepest slope? _____

b   Which have a negative slope? _____

Try to find all of the possible answers to each question.

c   Which cut the y-axis at $(0, 3)$? _____

d   Which has a y-intercept of $^-2$? _____

*e   Which is parallel to the x-axis? _____

(B) *Table it*

Complete this table. The first one is done for you.

Equation	$y = 4x + 3$	$y = 2x - 1$	$y = \frac{1}{2}x + 2$	$y = {}^-5x$	$y = 10 - 3x$	*	*
Gradient	4					3	$\frac{1}{4}$
Coordinates of y-intercept	$(0, 3)$					$(0, {}^-1)$	$(0, 0)$

(C) *Match it*

Match each line on the graph with an equation below.

a $y = 2x + 4$ _____
b $y = x - 2$ _____
c $y = \frac{-1}{2}x - 4$ _____
d $y = 2$ _____
e $y = \frac{-1}{2}x - 2$ _____
f $x = 2$ _____

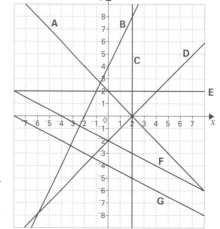

***(D)** *Write your own*

Write the equations of two lines which

a   have the same gradient as $y = \frac{1}{3}x + 5$. _____ _____

b   cut the y-axes at $(0, {}^-2)$. _____ _____

c   have a steeper slope than $y = 3x + 1$, but cut the y-axes in the same place. _____ _____

56 Reading Real-life Graphs

Let's look at ...
● estimating values from a graph

These are the points you need to know.

✓ We can use the graph to **estimate values**.

Example This graph shows the ferry charges for different lengths of cars.
We estimate the charge for a 420 cm car as £84.
We estimate the length of a car charged £115 as 575 cm

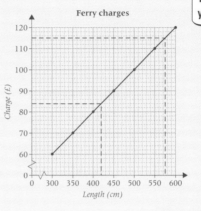

Ferry charges

Ⓐ *How hot is that?*

a It was 16°C in Cardiff and 68°F in Glasgow. Which city had the highest temperature? _____

b Estimate the missing values in this table.

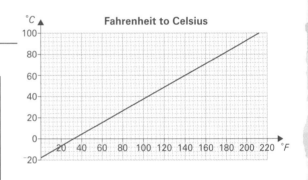

Fahrenheit to Celsius

Location	Average high temperature	
	°F	°C
London, UK	70°	
Bangkok, Thailand		35°
Kathmandu, Nepal		30°
Death Valley, USA.	116°	

c The highest recorded temperature in the Western Hemisphere was measured to be 134°F in 1913 in Death Valley. Estimate how many degrees Celsius this is. _____

d The record high temperature at the South Pole was 7·5°F, in December 1978. Estimate how many degrees celsius this is. _____

Ⓑ *Becky's first year*

This graph shows Becky's length and mass over the first year of her life. Estimate the answer to the following questions.

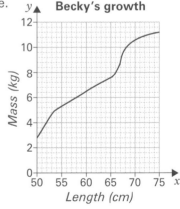

Becky's growth

a Becky was 50 cm long at birth. Estimate how much she weighed. _____

b On her first birthday Becky was 75 cm long. Estimate what her weight was. _____

c Complete this table.

Becky's length	Becky's mass
58 cm	
	8·4 kg
	5 kg

d About how much weight did Becky gain as her length increased from 60 cm to 65 cm? _____

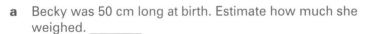

How did you find this? [**EASY**] [**OK**] [**HARD**]

57 Plotting Real-Life Graphs

Let's look at ...
- plotting real-life graphs and using them to answer questions

✓ Sometimes we **plot a real-life graph** using a **formula**.
  To do this we
  - decide how many points to plot
  - construct a table of values
  - choose suitable scales for the axes
  - plot the points accurately
  - give the graph a title and label the axes.

In some questions some of these are done for you.

These are the points you need to know.

✓ It is important to choose a **suitable scale** for the vertical axis or the graph can be **misleading**.

A School production

Connor is buying fabric to make costumes for the school play.
The fabric costs £2·50 per metre.

a Complete this table.

Length of fabric (*m*)	1	2	3	4	8
Cost (£)	£2·50	£5			

b Complete these coordinate pairs.
(1, 2·50), (2, 5), (3, _____), (4, _____), (8, _____)

c Draw a graph of this information.
Give your graph a title.
Draw a straight line through your points.

d Use your graph to estimate the following.
  i How much would 6 m of fabric cost? _____
  ii How much would 3·5 m of fabric cost? _____
  iii How much fabric could Connor get for £17·50? _____

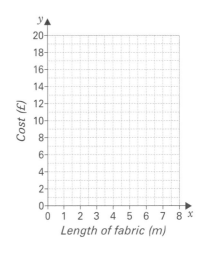

B Super Savers

Two friends, Jordan and Michael, each start new weekend jobs at the beginning of the year.

  Jordan mows lawns and earns £40 per month.
  Michael washes windows and earns £30 per month.

They each put all of their earnings into their savings accounts.

a Jordan starts the year with £20 in his account, and Michael with £50 in his account.
Complete these tables.

JORDAN

Number of months	0	1	2	3	4
Jordan's savings (£)	20				

MICHAEL

Number of months	0	1	2	3	4
Michael's savings (£)	50				

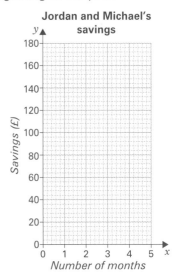

Jordan and Michael's savings

b Draw the graphs of each boy's savings on this grid.
Join each set of points with a straight line.
Label each line.

c When do the boys have the same amount of money in their accounts? _____

d After 4 months which boy has the most money saved? _____

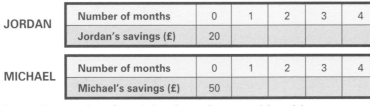

58 Distance/Time Graphs

Let's look at ...
● interpreting and drawing distance/time graphs

✓ This shows a **distance/time graph**.

Lucy rowed a dinghy at a steady pace. She rowed 6 km in 3 hours. She rested for an hour then rowed back to the start in 2 hours.
Lucy drew this graph to show the relationship between distance and time.

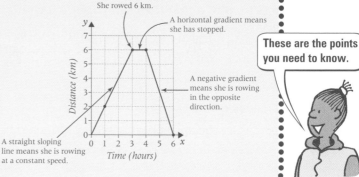

These are the points you need to know.

She rowed 6 km.

A horizontal gradient means she has stopped.

A negative gradient means she is rowing in the opposite direction.

A straight sloping line means she is rowing at a constant speed.

(A) Building supplies

Jack and his father took 10 minutes to drive 4 km. They stopped for 5 minutes while Jack bought some wood and nails. They then drove for 15 minutes to the paint shop, which is 12 km from home. Jack spent another 5 minutes in this shop. They then drove home in 20 minutes.

Finish the graph for Jack's drive.

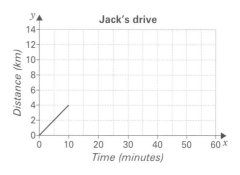

Jack's drive

(B) Tricycle adventure

Olivia's adventure

On her third birthday Olivia set off on an adventure. She rode her tricycle 50 metres to the front gate. This took 1 minute. She then spent another 1 minute getting the gate open. With a spurt of energy she spent the next minute cycling a further 100 metres to an interesting post box. Olivia stared at this for a full 2 minutes. Finally she cycled 150 metres further to her Grandparents' home. This took her 5 minutes.

Draw a distance/time graph for Olivia's adventure.

(C) To boarding school

This graph shows Freddy's train journey from home to boarding school.

a   How far away is Freddy's school? _____

b   How long did Freddy's journey take? _____

c   For how long in total did Freddy stop during his journey? _____

d   How far did he travel in the first 2 hours? _____

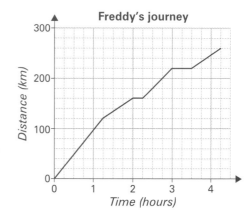

Freddy's journey

How did you find this?   **EASY**   **OK**   **HARD**

59 Interpreting Real-life Graphs

Let's look at ...
● **describing trends from the shape of a graph**

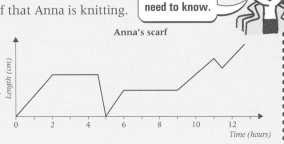

These are the points you need to know.

✓ To help **interpret a graph** we often look at the shape of it.

Example   This graph shows the length of a scarf that Anna is knitting.
The graph shows that Anna knitted
for 2 hours, then stopped knitting.
Almost 3 hours later the cat found
her knitting and destroyed it all!
Anna then started her knitting again,
rested, and knitted some more.
She unpicked a small amount, and
continued to knit straight away.

A What was dropped?

Which of these matches the graph shown?

A   Dropping a plastic cup.

B   Dropping an egg.

C   Dropping a paper aeroplane.

D   Dropping a rubber ball.

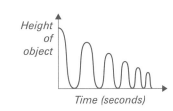

B What's it mean?

For each of these graphs, explain what you can tell from the shape of the graph.
Mention trends and how the variables are changing.

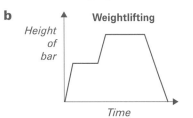

a _____
_____

b _____
_____

c _____
_____

C Who is the tallest?

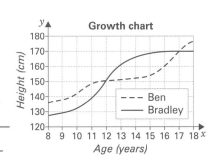

Ben's and Bradley's heights are plotted on this graph.

a   Who kept growing for longer? _____

b   Who is the shortest adult? _____ How tall is he? _____

c   At what ages were Ben and Bradley the same height? _____

d   Who grew the most between the ages of 12 and 14? _____

60 Sketching and Interpreting Real-life Graphs

Let's look at ...
● sketching line graphs for real-life situations

 Sometimes we **sketch a graph** for a real-life situation or **interpret a sketch**.

Example Beakers A and B are filled with boiling water.
Beaker A is left in the room.
Beaker B is put in the fridge.
We can sketch a graph of temperature against time for the two beakers.

These are the points you need to know.

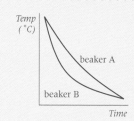

A *Flying high*

Match each of these with one of the graphs below.

a radio controlled aeroplane ____   **b** swing in a playground ____

c sky rocket ____   **d** model hot air balloon taking off and catching fire ____

B *Breakfast time*

A factory fills two sizes of box with the same steady flow of breakfast cereal.

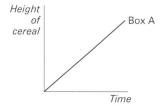

The line graph showing the height of cereal in Box A against the time is shown.
Sketch the line graph for Box B on the same set of axes.

*C *There's a hole in my bucket*

Three buckets each have a hole in them.
Apart from the different holes the buckets are identical.
All three buckets are filled with water.
The line graph for the depth of water in bucket 1 over time has been drawn.
Sketch the graphs for the depth of water in buckets 2 and 3.

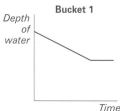 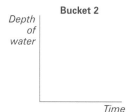 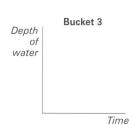

How did you find this? EASY   OK   HARD

SHAPE, SPACE AND MEASURES

61 Angles Made with Intersecting and Parallel Lines

Let's look at ...
● complementary and supplementary angles
● corresponding and alternate angles

✓ A pair of angles that add to **90°** are called **complementary angles**.
A pair of angles that add to **180°** are called **supplementary angles**.

✓ Angles made with parallel lines.

$c = d$
corresponding angles are equal

$f = g$
alternate angles are equal

These are the points you need to know.

A) Sort them

Sort each pair of angles into the correct box.

A 46° and 44°   **B** 32° and 148°

C 52° and 38°   **D** 17° and 83°

E 101° and 79°   **F** 83° and 87°

G 24° and 166°   **H** 32° and 58°

I 125° and 55°

Complementary angles	Supplementary angles	Neither
A		

B) Find d

Find the size of angle d. Give a reason.

a

$d =$ _____
_____
_____
_____

b

$d =$ _____
_____
_____
_____

C) Name that angle

For each diagram name

		Diagram 1	Diagram 2
a	2 pairs of corresponding angles	h and d,	
b	2 pairs of alternate angles		
c	2 angles that are equal to angle a		
d	4 pairs of supplementary angles		

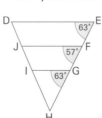

Diagram 1

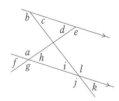

Diagram 2

*D) Are they parallel?

For each diagram name a pair of parallel lines. Explain how you know.

a

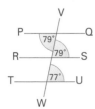

____ and ____ are parallel because
_____
_____
_____

b

____ and ____ are parallel because
_____
_____
_____

How did you find this?   EASY   OK   HARD

62 Using Geometric Reasoning to Find Angles

Let's look at ...
● proving that an angle has the value shown
● using geometric reasoning to find the size of missing angles

✓ Sometimes it is not obvious how to find the size of an angle. It helps to do these:

1 Write down what you know.

2 Work out any angles that it looks like you might need.

3 Use what you know from **1** and **2**. Show your working and always give reasons.

> These are the points you need to know.

Example   Prove that $e = 38°$.

e is an alternate angle to f.

$f + 142° = 180°$ angles on a straight line add to 180°

$f = 180° - 142° = 38°$

So $e = 38°$ alternate angles on parallel lines are equal

> There is often more than one way to find the answer.

✓ You will need to recall these properties of angles.

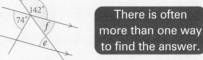

vertically opposite angles are equal	angles on a straight line add to 180°	angles at a point add to 360°	interior angles of a triangle add to 180°
$a = b$	$x + y + z = 180°$	$c + d + e = 360°$	$a + b + c = 180°$

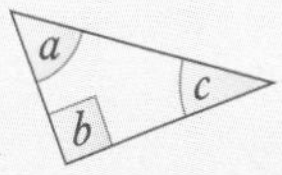

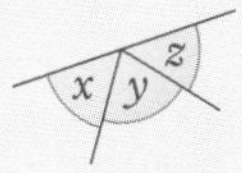

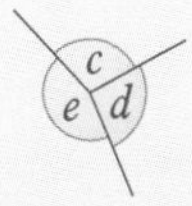

> Look at page 000 for two more properties.

A Prove it

Prove that p has the value shown. Show your working clearly and give reasons.

a

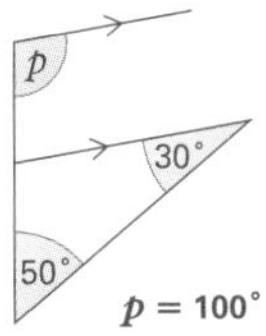

$p = 100°$

b

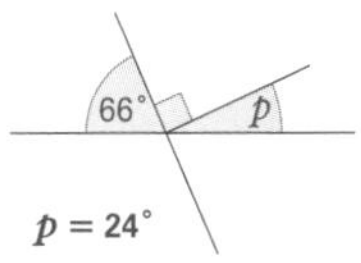

$p = 24°$

B Find it

Find the values of a, b, c, d, e and f. Show your working clearly and give reasons.

a

b

C Solve it

Calculate the value of x. You will need to write and solve an equation.

a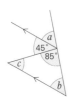

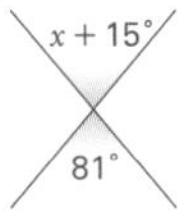

b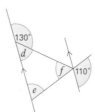

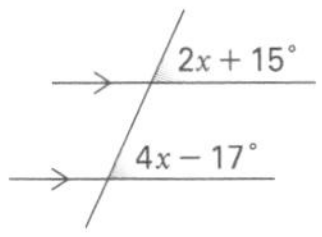

63 Angles in Triangles

Let's look at ...
● the relationship between exterior and interior angles in triangles

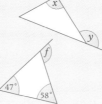

These are the points you need to know.

✓ Angle x is an **interior** angle of the triangle.
  Angle y is an **exterior** angle of the triangle.

✓ **The exterior angle of a triangle is equal to the sum of the two opposite interior angles**.

  Example $f = 47° + 58°$
  $\qquad\ \ = 105°$

Exterior angles are outside the shape.

A Quick questions

Find the value x.

a   **b**   **c**   **d**    **e**

$x =$ _____ $x =$ _____ $x =$ _____ $x =$ _____ $x =$ _____

B What's the reason?

Find the value of x and y. Give reasons.

a  $x =$ _____ _____

_____

_____

$y =$ _____ _____

_____

b $x =$ _____ _____

_____

_____

$y =$ _____ _____

_____

C Hex-angles

Each exterior angle of a regular hexagon is 60°.

Find the size of angle x. Give reasons. _____

_____

_____

D Tricky triangles

Find the size of the angles marked with letters.
Write down all of your working.

a 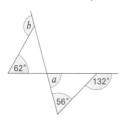   **b** 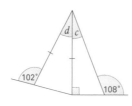   **c**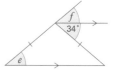

$a =$ _____ _____ $c =$ _____ _____ $e =$ _____ _____

_____   _____   _____

_____   _____   _____

$b =$ _____ _____ $d =$ _____ _____ $f =$ _____ _____

_____   _____   _____

How did you find this?   EASY   OK   HARD

64 Angles in Quadrilaterals

Let's look at ...
● the sum of the interior angles of a quadrilateral

 The **sum of the interior angles of a quadrilateral** add to **360°**.

$a + b + c + d = 360°$

Example   We can find the value of the angle marked with letters.
● $a = 180° - 82°$ (angles on a straight line add to 180°)
    $= 98°$
● $b = 82°$ (alternate angles on parallel lines are equal)
● $c + 98° + 82° + 71° = 360°$ (angles in a quadrilateral add to 360°)
    $c = 360° - 98° - 82° - 71°$
    $c = 109°$

There is often more than one way to find the answer.

These are the points you need to know.

A One quad at a time

Find the size of the angles marked with letters.

 $a = $ _____

$b = $ _____

$c = $ _____

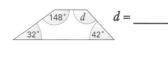 $d = $ _____

 $e = $ _____

arrowhead

 $f = $ _____

kite

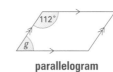 $g = $ _____

parallelogram

B Double trouble

Find the value of the angles marked with letters. Show your working and give reasons.

a

Label any other angles you need. Example: You could name this angle c.

b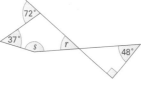

$p = $ _____ _____
_____
_____
$q = $ _____ _____
_____

$r = $ _____ _____
_____
_____
$s = $ _____ _____
_____

*C Prove it

This diagram shows two overlapping squares and a straight line.
Prove that $x = 125°$ and $y = 55°$.

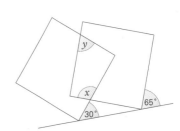

_____
_____
_____
_____

65 Visualising and Sketching 2-D Shapes

Let's look at ...
● imagining 2-D shapes

These are the points you need to know.

✓ We can **visualise 2-D shapes**.

Example Imagine a rectangle, cut along the diagonal to make two triangles. The triangles are put together along sides of equal length. These shapes can be made.

Can you think of any other ways to put the triangles together?

You can rotate or flip the triangles.

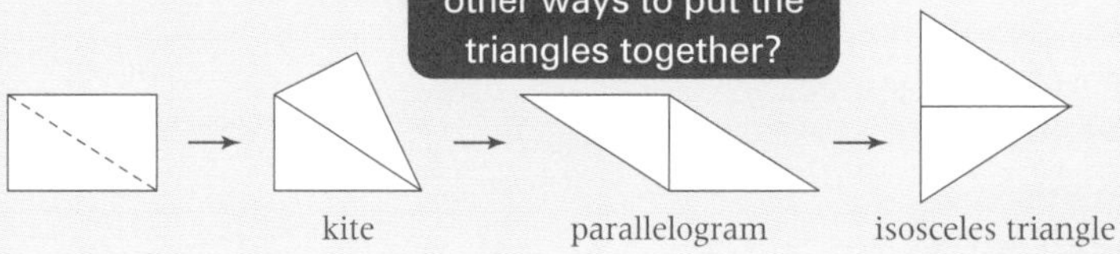

kite → parallelogram → isosceles triangle

A Imagine!

a i Imagine a square piece of paper is cut into quarters. Each quarter is the same size and shape. Draw and name the possible shapes that can be made.

You could make right-angled triangles.

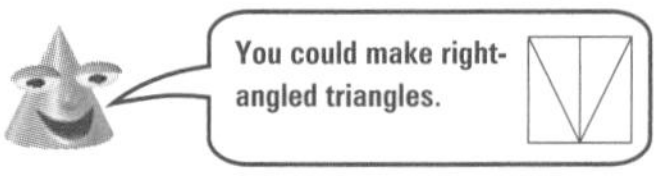

ii Madison cut her square piece of paper into four right-angled triangles. She put the triangles together along sides of equal length and made a new shape. Draw and name the shapes she could have made.

b Imagine two identical isoceles trapeziums put together along sides of equal length. Draw and name the shapes that can be made.

An isoceles trapezium has 2 equal sides.

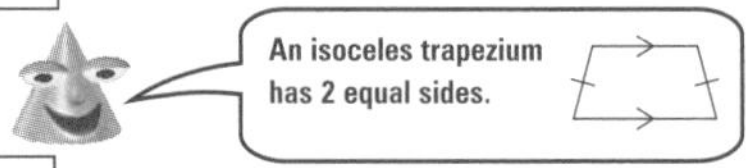

c Imagine a shape is cut in half to make two identical scalene triangles. The two halves are put together along matching edges to make a new shape. Draw and name the possibilities.

B Describe!

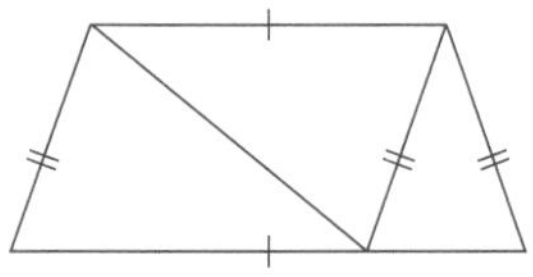

Describe this shape using as much detail as possible.

_____

_____

_____

_____

_____

How did you find this?   EASY   OK   HARD

66 Properties of Triangles and Quadrilaterals

Let's look at ...
● side, angle, diagonal and symmetry properties of triangles and quadrilaterals

✓ The **symmetry properties** of some **special triangles and quadrilaterals** are summarised in this table.

Remember if a shape has rotation symmetry of order 1, it has no rotation symmetry.

These are the points you need to know.

Shape	Number of lines of symmetry	Order of rotation symmetry
Equilateral △	3	3
Isosceles △	1	1
Square	4	4
Rectangle	2	2
Parallelogram	0	2
Rhombus	2	2
Kite	1	1
Isosceles trapezium	1	1
Trapezium	0	1
Arrowhead	1	1

We can use these symmetry properties to work out other properties of the shapes.

(A) What am I?

a I have 3 sides.
I have 1 line of symmetry.

What shape am I?

b I have 2 lines of symmetry.
I have rotation symmetry of order 2.
All of my sides are equal.
What shape am I?

*c I have 1 line of symmetry.
I have no rotation symmetry.
My diagonals cross at right angles.
What two shapes could I be?

_____   _____   _____

(B) True or false?

Write true or false for each of these.

 You could work out the answers to these using the symmetry properties of the shapes.

a A rectangle has opposite sides and opposite angles equal. _____

b A parallelogram has diagonals that are equal. _____

c The diagonals of a square cross at right angles. _____

d An isosceles trapezium has diagonals that bisect the angles. _____

Write true or false for each of these and explain why.

e All rhombuses are parallelograms. _____

f All rhombuses are squares. _____

(C) Angles and sides

Use the properties of each shape to find angle X and the length of XY. Explain your reasoning clearly.

a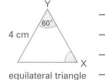
equilateral triangle

_____
_____
_____
_____

b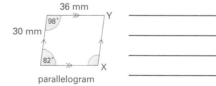
parallelogram

_____
_____
_____
_____

*(D) What's in common?

What do the shapes in each of these lists have in common?

a Parallelogram, square, rhombus, rectangle **4 sides and** _____

b Isosceles triangle, kite, arrowhead, isosceles trapezium _____

67 Tessellations

Let's look at ...
● identifying shapes that tessellate

These are the points you need to know.

✓ A shape will **tessellate** if it can be used to completely fill a space with no overlapping and no gaps.

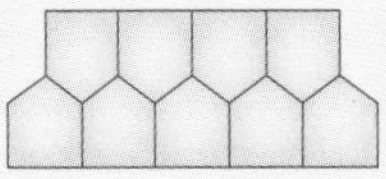

A tessllation can be made by reflecting, rotating or translating a shape.

A Which will work?

Decide whether or not each of these shapes will tessellate.
Sketch a picture to show this. The first one is done for you.

a kite

Yes

b rectangle

_____

c circle

_____

d rhombus

_____

e isosceles trapezium

_____

*f regular octagon

_____

B Regular reasoning

Patrick wrote this report about equilateral triangles.

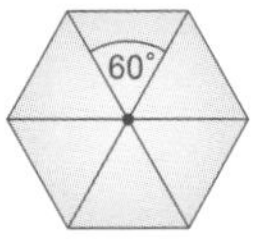

60°

● Equilateral triangles tessellate.

● The internal angle of an equilateral triangle is 60°.
This means that exactly 6 triangles will meet at a point.

There are 360° at a point.
360° ÷ 60° = 6

a Which of these regular polygons will tessellate? _____

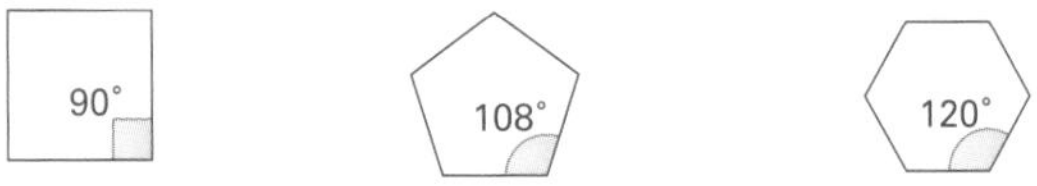

90°

square

108°

pentagon

120°

hexagon

b Use the internal angles of the shapes to explain how you know this. _____
_____
_____

C Nelly's house

a Nelly wants to lay cobblestones of this shape in her driveway.
Draw two different patterns that she could create.

There can be no overlapping and no gaps.

*b Nelly uses octagonal and square tiles on her kitchen floor.
Draw the pattern she creates.

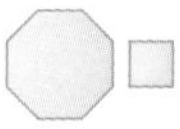

There can be no overlapping and no gaps.

How did you find this? 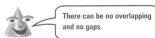 EASY OK HARD

68 Congruence

Let's look at ...
● identifying congruent shapes

These are the points you need to know.

✓ **Congruent** shapes are exactly the same shape and size.
In congruent shapes, corresponding sides and corresponding angles are equal.

Example

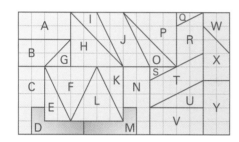

LM = RT	∠M = ∠T
MN = TS	∠N = ∠S
LN = RS	∠L = ∠R

These shapes are congruent.

A Hidden shapes

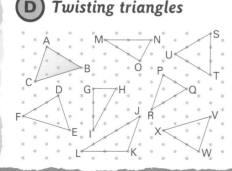

a Name all of the congruent shapes in this diagram.

D and M _____ _____

_____ _____

_____ _____

_____ _____

_____ _____

b Which three shapes are not congruent with any other shape? _____, _____, _____

B Yay or nay?

Decide whether these pairs of shapes are congruent. Explain why or why not.

a

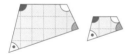

b

_____

_____

_____

_____

C Back track

Each of these shapes was made by joining two congruent shapes together.
Divide each shape in two to show the original two shapes and name them.
The first one is done for you.

a

right-angled triangles

b

_____

c

_____

d

rhombus

_____

D Twisting triangles

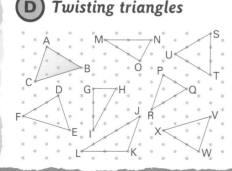

a Name all of the triangles that are congruent to triangle ABC.

_____

b For each triangle congruent to ABC, name the angle equal to angle B. _____

c For each triangle congruent to ABC, name the side equal to BC. _____

How did you find this? **EASY** **OK** **HARD**

69 Visualising and Sketching 3-D Shapes and Nets

Let's look at ...
● visualising faces and edges on 3-D shapes
● using isometric paper to draw 3-D shapes

✓ We often draw 3-D shapes using **isometric** (triangle dotty) paper.
An isometric drawing has vertical edges drawn as vertical lines, horizontal edges drawn at 30° to the horizontal.
When we make an isometric drawing, some edges are not shown.
The dashed lines show the hidden edges.

These are the points you need to know.

A Door stop

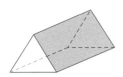

Danielle made this wooden door stop.
She painted *one* triangular face white, and the other *four* faces purple.

a How many edges does the door stop have? _____

b At how many edges do a white and purple face meet? _____

c At how many edges do two purple faces meet? _____

B Use the dots

a Make an isometric drawing of these shapes

i   ii   iii

b Use dashed lines to show the hidden edges in shape **i**.

C Pens and pencils

Harrison is making a box for his stationery.
He wants the box to have long stripes on the top and bottom.
He wants to have dots around the bottom of each side.
Complete the pattern on these nets so that they will fold to make Harrison's box.

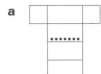

a 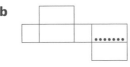   **b**   **c**

D Puzzle

Here are three views of the same cube.

Which letters are opposite each other? _____, _____, _____

How did you find this?   EASY   OK   HARD

70 Plans and Elevations

Let's look at ...
● plan views, front elevations and side elevations

✓ The view from the top of a shape is called the **plan view**.

✓ The view from the front is called the **front elevation**.

✓ The view from the side is called the **side elevation**.

Example

These are the points you need to know.

A Match them up

Choose from the box below to match each 3-D shape with its front, side and plan view.

a

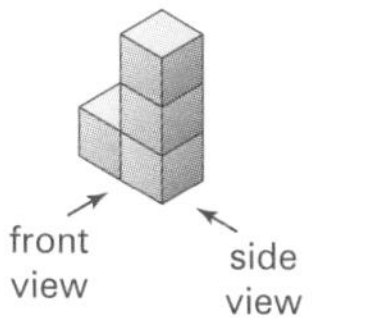

front view   side view

front elev. __F__

side elev. ____

plan view ____

b

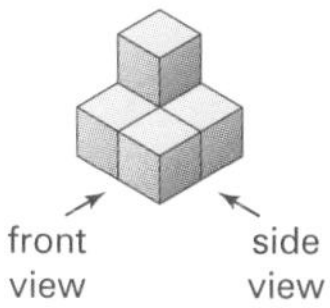

front view   side view

front elev. ____

side elev. ____

plan view ____

c

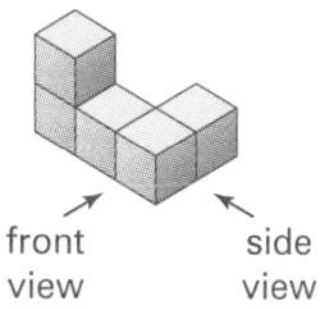

front view   side view

front elev. ____

side elev. ____

plan view ____

d

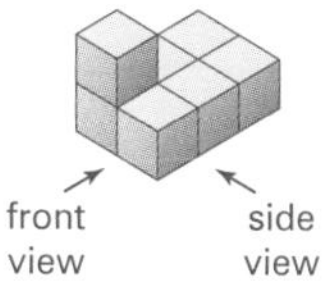

front view   side view

front elev. ____

side elev. ____

plan view ____

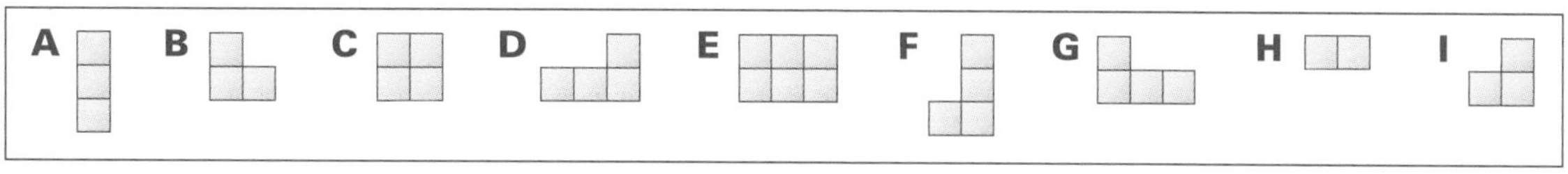

| A | B | C | D | E | F | G | H | I |

B What's your view?

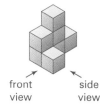

front view   side view

Draw the front elevation, plan view and side elevation of this shape.

front elev.

plan view

side elev.

C Dotty diagrams

The diagram on the left represents a plan view of the solid on the right.

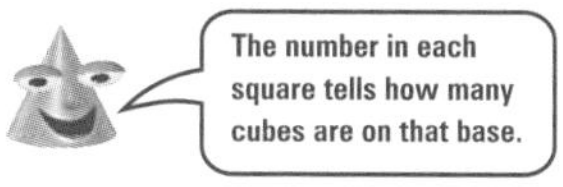

The number in each square tells how many cubes are on that base.

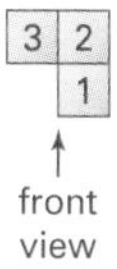

| 3 | 2 |
| 1 | |

↑ front view

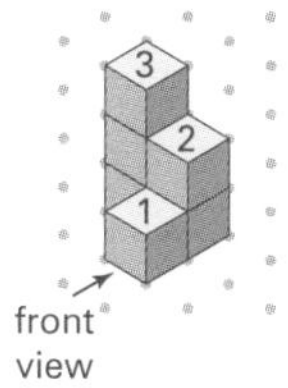

front view

Draw the 3-D shape that this diagram represents.

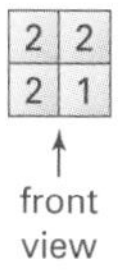

| 2 | 2 |
| 2 | 1 |

↑ front view

* D Your choice

Shape A

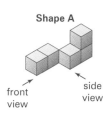

front view   side view

Draw two other shapes with the same side view and two other shapes with the same plan view as Shape A.

71 Construction

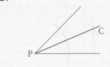

Let's look at ...
● constructing mid-points and perpendiculars of line segments
● constructing bisectors of angles

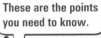
These are the points you need to know.

✓ We use a compass and a ruler to make the following **constructions**.

● mid-point (R) and perpendicular bisector (PQ) of a line segment (BC)

● bisector (PC) of an angle (P)

● perpendicular (AR) from a point (A) to a line (BC)

● perpendicular (QR) from a point (P) on a line (BC)

A *Perpendiculars*

Leave your construction lines.

a Construct the perpendicular bisector of the line segment DE.

b Construct the perpendicular from point P on the line FG.

B *Draw the design*

Find the mid-points by constructing perpendicular bisectors.

Complete larger versions of these designs by following the instructions.

a

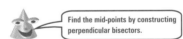

i Find the mid-point of RS. Label this A.
ii Find the mid-point of ST. Label this B.
iii Find the mid-point of RT. Label this C.
iv Use your ruler to join points A, B and C.

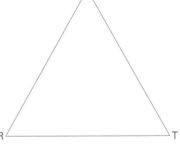

b

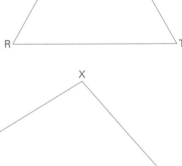

i Construct the perpendicular from point X to the line YZ. Label as W the point where this perpendicular meets YZ.
ii Construct the bisector of angle XWZ.
iii Construct the bisector of angle XWY.
iv Measure the angle between the lines you constructed in **i** and **ii**. _____

How did you find this? **EASY** **OK** **HARD**

72 Constructing Triangles and Quadrilaterals

Let's look at ...
- constructing triangles using a ruler and a protractor or a compass
- constructing quadrilaterals using a ruler and a protractor or a compass

✓ We can **construct a triangle** using a ruler and protractor if we are given two sides and the included angle (SAS) or two angles and the side between them (ASA).

✓ If the three sides (SSS) of a triangle are given we can construct the triangle using a ruler and compass.

✓ Quadrilaterals can be constructed in a similar way.

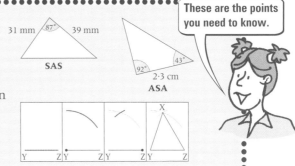

These are the points you need to know.

A SAS and ASA

Construct triangle ABC where

a AB = 3·4 cm, AC = 2·9 cm, ∠A = 130°
Measure the length of BC _____

b AB = 4·2 cm, ∠A = 71°, ∠B = 55°
Measure angle C _____

B SSS

Draw triangles with sides of these lengths.
Measure the angle between the first two sides given.

a 3 cm, 4 cm, 6 cm
Angle = _____

b 2 cm, 3·5 cm, 5 cm
Angle = _____

C Quads

Construct these quadrilaterals which have been sketched. Measure the dashed length and angle a on your construction.

a

b
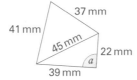

Dashed length = _____
angle a = _____

73 Loci

Let's look at ...
● drawing and describing loci

These are the points you need to know.

✓ A **locus** is a set of points that satisfy a rule or set of rules.

Example   The locus of a robot moving so that it is always the same distance from a fixed point is a circle.

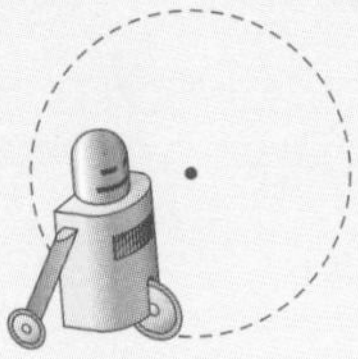

The plural of locus is loci.

A *Riding School*

a   Kylie rode a horse which was roped to a post in the middle of the field. The horse walked so that it was always the same distance from the post. Describe Kylie's locus in words.

_____

b   Nick rode his horse between two parallel fences (shown by GH and IJ). the horse canters through, so that it is always the same distance from the fences either side. Draw Nick's locus on the diagram. Describe Nick's locus in words. _____

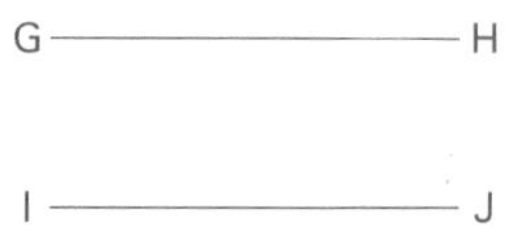

G ———————— H

I ———————— J

_____

c   Kate rode her horse in a field with two large trees (shown by T1 and T2). Kate rode so that she was never closer to one tree than the other. Draw Kate's locus on the diagram. Describe Kate's locus in words. _____

T1   T2
•   •

_____

d   R•———— S   Allan mounted his horse in the corner of a fenced field (point R). He rode so that he stayed equidistant from the two side fences (shown by QR and RS). Draw and describe Allan's locus. _____

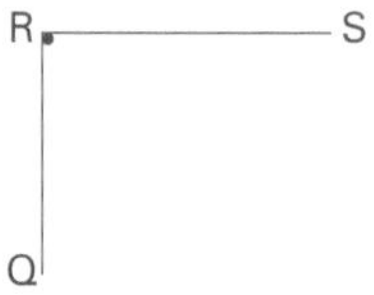

Q•

_____

*e   Charles rode his horse around a water trough (shown by WX). Charles kept the horse walking exactly 2 m away from the trough. Sketch this locus. Put measurements on your sketch.

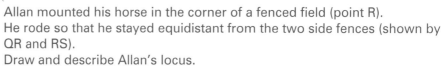

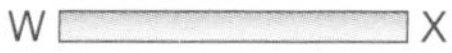

W ▭ X

74 Coordinates and Transformations

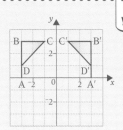

Let's look at ...
● giving the coordinates of an image after reflection, rotation or translation

These are the points you need to know.

✓ We can **reflect, rotate or translate** shapes.

Example The shape ABCD has been reflected in the y-axis, to give the image A'B'C'D'.
The coordinates of the vertices of the original shape are $(^-3, 0)$, $(^-3, 3)$, $(^-1, 3)$ and $(^-3, 1)$.
The coordinates of the vertices of the image are $(3, 0)$, $(3, 3)$, $(1, 3)$ and $(3, 1)$.

A *What a transformation*

1 a Write down the coordinates of A, B, C and D.
A = _____   B = _____   C = _____   D = _____

b Write down the coordinates of A', B', C' and D' after
  i reflection in the x-axis.
   A' = _____   B' = _____   C' = _____   D' = _____

  ii reflection in the y-axis.
   A' = _____   B' = _____   C' = _____   D' = _____

  iii rotation 180° about the origin.
   A' = _____   B' = _____   C' = _____   D' = _____

Start with the original coordinates each time.

2

a Three of the vertices of a square are $(3, 3)$, $(^-3, 2)$ and $(^-2, ^-4)$. Plot these points.

b Write down the coordinates of the fourth vertex. _____

c Write down the coordinates of the four vertices after
  i translation 1 unit to the right and 3 units up.
  _____

  ii rotation 90° about the origin
  _____

  iii reflection in the y-axis
  _____

Start with the original coordinates each time.

3 a Write down the coordinates of the vertices of this shape.

b Write down the coordinates of the vertices after
  i translation 1 unit to the left and 2 units up
  _____

Start with the original coordinates each time.

  ii rotation 270° about the origin
  _____

  ＊iii reflection in the line $y = ^-1$
  _____

Remember to rotate anticlockwise.

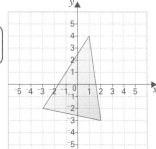

75 Combinations of Transformations

Let's look at ...
● transforming shapes with two or more transformations

These are the points you need to know.

✓ We can transform shapes with a **combination of transformations**.

Example   A has been reflected in the *y*-axis and then translated 3 units down to get B.

✓ Some **combinations** have a predictable result.
Reflection in two parallel lines is equivalent to translation.
Reflection in two perpendicular lines is equivalent to a half-turn rotation.
Two rotations about the same centre are equivalent to a single rotation.
Two translations are equivalent to a single translation.

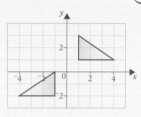

A Reflect and reflect

1 a Reflect each shape in m_1 and then in m_2.

b Circle the single transformation which is equivalent to reflection in two parallel lines.
a reflection   **a rotation**   **a translation**

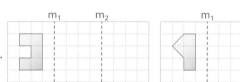

2

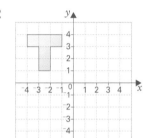 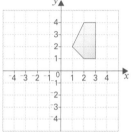

a For each shape given
  i reflect the shape in the *y*-axis.
  ii reflect this image in the *x*-axis.

b Circle the single transformation which is equivalent to reflection in two perpendicular lines.
a full-turn rotation   **a half-turn rotation**
a translation

B Translate and translate

a Translate this triangle 4 units right and 5 units down.

b Translate the image you get 3 units left and 2 units up.

c What single translation is equivalent to the two translations in **a** and **b**? _____

_____

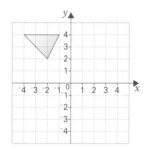

C Rotate and rotate

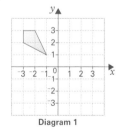

Diagram 1

a Use diagram 1. Rotate the shape 180° about the origin. Then rotate the image 90° about the origin.

b What single rotation is equivalent to these two rotations? _____° rotation about _____

*c Use diagram 2. Rotate the shape 90° about (1, 1). Then rotate the image 180° about (1, 1).

*d What single rotation is equivalent to these two rotations? _____° rotation about _____

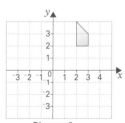

Diagram 2

76 Symmetry

Let's look at ...
● describing the reflection and rotation symmetry of a shape

These are the points you need to know.

✓ A shape has **reflection symmetry** if one half of the shape can be reflected in a line to the other half. The line is a **line of symmetry**.

✓ A shape has **rotation symmetry** if it fits onto itself **more than once** during a complete turn.
The **order of rotation symmetry** is the number of times a shape fits exactly onto itself during one complete turn.

This shape has 3 lines of symmetry and rotation symmetry of order 3.

✓ If a shape has rotation symmetry of order 1, we say it does not have rotation symmetry.

(A) Shapes and signs

a   b   c   d   e

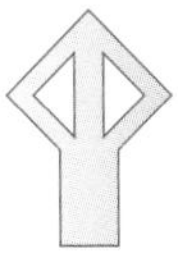

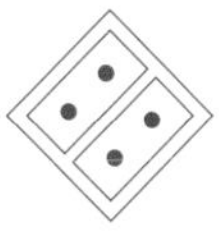

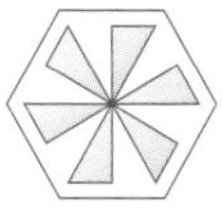

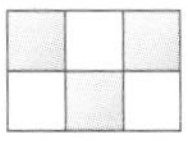

Complete this table.

	a	b	c	d	e
Number of lines of symmetry					
Order of rotation symmetry					

(B) Show it's true

Peter said, '*The base angles of an isoceles triangle are equal*'.
His teacher asked him to show that this is true using the symmetry properties of the shape.
Peter said, '*If I fold an isoceles triangle along its line of symmetry, the base angles fit exactly on top of one another*'.
Explain how you could show that each of these is true.

a The angles of a regular hexagon are all equal. _____

_____

b The diagonals of a rectangle bisect each other. _____

_____

(C) Crossword time

Complete this crossword grid so that it has rotation symmetry of order 4.

*(D) Jigsaw

Cathy has these puzzle pieces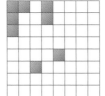
Show how she can put them together to make a shape with

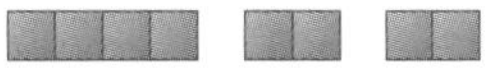

a	b	c
1 line of symmetry and no rotation symmetry	2 lines of symmetry and rotation symmetry of order 2	0 lines of symmetry and rotation symmetry of order 2

77 Enlargement

Let's look at ...
● scale factors and centres of enlargement

✓ To **enlarge** a shape we know the **scale factor** and the **centre of enlargement**.

Example PQR has been enlarged by scale factor **2** and centre of enlargement X.
Each point on the image, P′Q′R′, is **2** times as far from X as the same point on PQR.
Each length on P′Q′R′ is twice as long as the corresponding length on PQR.

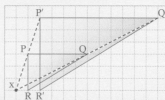

These are the points you need to know.

P′Q′R′ was plotted by drawing a dashed line from X through each point P, Q and R, and extending the dashed line to make it twice as long.

A Find the factor

Shape A has been enlarged to shape B. What is the scale factor of the enlargement?

a
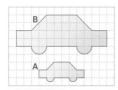
scale factor = _____

b
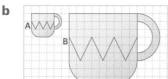
scale factor = _____

c
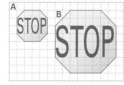
scale factor = _____

B Lamp light

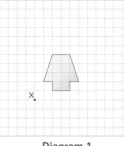

Diagram 1

a Enlarge the lamp in diagram 1 by a scale factor of 2. Use X as the centre of enlargement.

b Enlarge the lamp in diagram 2 by a scale factor of 3. Use Y as the centre of enlargement.

c | angles   lengths |

Use these words to complete this sentence.
When a shape is enlarged the _____ stay the same but the _____ change.

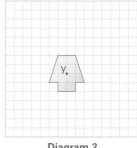

Diagram 2

C What are the points?

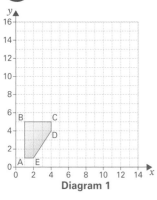

Diagram 1

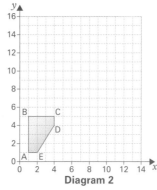

Diagram 2

a Write down the coordinates of ABCDE.

_____

b Enlarge ABCDE on diagram 1 by a scale factor of 2. Use the origin as the centre of enlargement.
Write down the coordinate of A′B′C′D′E′.

_____

c Enlarge ABCDE on diagram 2 by a scale factor of 3. Use the origin as the centre of enlargement.
Write down the coordinate of A′B′C′D′E′.

_____

d What do you notice about the coordinates of ABCDE and the coordinates of its images? ____

_____

How did you find this? [EASY] [OK] [HARD]

78 Scale Drawing

Let's look at ...
● interpreting and drawing scaledrawings

These are the points you need to know.

✓ A **scale drawing** represents something in real life.

Example This is a scale drawing of a car.
Each millimetre on the drawing represents 6 cm in real life. So 1 mm on the drawing represents 60 mm in real life.
The car is 50 mm on the drawing.
In real life it is 50 × 60 = 3000 mm = 300 cm = 3 m

scale 1 mm represents 6 cm

✓ To make a scale drawing you need to
1 make a rough sketch with the real life measurements written on it
2 choose a scale
3 work out what each measurement on the scale drawing will be.

(A) *Mediterranean holiday*

The man is 1·8 m tall.

Estimate these.

a the height of the picnic table _____
b the height of the palm tree _____
c the height of the boat _____
d the length of the fence _____

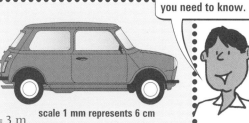

(B) *Katie's house*

This is an accurate floor plan of the ground floor of Katie's house. The scale is 1 cm represents 150 cm. What are the actual dimensions of the

a living room _____

b kitchen _____

c hallway _____

The dimensions are the length and the width.

(C) *Map work*

a On a street map 1 cm represents 300 m. What is the actual length in metres, of these streets measured on the map?
i Royal Cres 3 cm _____ **ii** Trinity Rd 6 cm _____ **iii** High St 12 cm _____

b On a map of a sport's stadium 1 mm represents 50 cm. On the map the swimming pool is 50 mm long. What is the real life length of the pool **i** in cm _____ **ii** in m? _____

c Tanya's map of a ski resort shows the hire shop being 2 cm long. In real life the hire shop is 20 m long. What is the scale of this map? 1 cm represents _____

d Sanjay's map of an island shows a road which is 6 cm long. It is actually 3 km long. What is the scale of this map? _____

(D) *Philip's backyard*

Make an accurate scale drawing of this sketch.
Use the scale 1 cm represents 2 m.

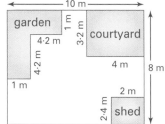

79 Finding the Mid-point of a Line

Let's look at ...
● finding the coordinates of the mid-point of a line segment

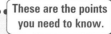

These are the points you need to know.

✓ The **mid-point** of the line segment joining $A(x_1, y_1)$ to $B(x_2, y_2)$ is given by $(\frac{x_1 + x_2}{2}, \frac{y_1 + y_2}{2})$.

Example The mid-point, M, of the line joining AB is $(\frac{3 + 13}{2}, \frac{4 + 10}{2}) = (8, 7)$.

$(x_1, y_1) = (3, 4)$
$(x_2, y_2) = (13, 10)$

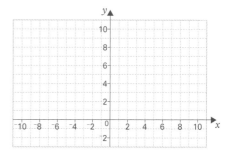

A Horizontals

Find the coordinates of the mid-point of DE.
Plot D, E and the mid-point to check you are correct.

a D(1, 3), E(7, 3) ____   **b** D(2, 5), E(8, 5) ____

c D(⁻4, 6), E(2, 6) ____   **d** D(⁻3, 8), E(6, 8) ____

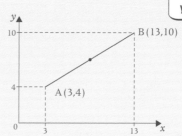

B Verticals

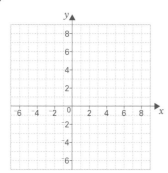

Find the coordinates of the mid-point of FG.
Show your working.
Plot F, G and the mid-point to show you are correct.

a **F**(4, 1) **G**(4, 7) _____

b **F**(6, ⁻5) **G**(6, 2) _____

C Diagonals

You will need some extra paper for the working for **c**, **d**, **e** and **f**.

a Find the mid-point of the line segment MN.
Show your working in the box.
Plot M, N and the mid-point to check that you have the correct answer.

i M(1, 4), N(5, 8) _____

ii M(⁻2, 1), N(⁻5, 5) _____

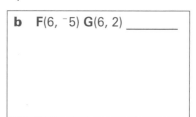

∗b The vertices of a rectangle are (⁻5, ⁻2), (⁻5, ⁻7), (7, ⁻7), (7, ⁻2).
Find the mid-point of a diagonal of this rectangle.

_____

How did you find this?   EASY   OK   HARD

80 Metric Conversions

Let's look at ...
● converting length, mass, capacity, area and time units

✓ You need to know these **metric conversions**.

These are the points you need to know.

mass	capacity	area
1 tonne = 1000 kg	$1\,\ell = 1000\ cm^3$	1 hectare (ha) = 10 000 m^2
	$1\ m\ell = 1\ cm^3$	
	$1000\ \ell = 1\ m^3$	

Examples $52\,800\ m^2 = (52\,800 \div 10\,000)ha$ $3000\ cm^3 = 3\,\ell$ or $4680\,\ell = (4680 \div 1000)m^3$
  = **5·28 ha**   = **3000 mℓ**   = **4·68 m^3**

A *Earth is slowing down*

<u>I</u> **2·6**	4500	8	2340	72 000	4·5	0·08	<u>I</u> **2·6**	0·072	0·072	<u>I</u> **2·6**	0·4	4500

2·34	72 000	8 ,	80 000	0·8234	260	8000	72 000	80 000	72 000

4·5	0·4	4500	260	80	72 000	8	0·072	72 000	8	0·045

2·34	72 000	8	80 000 .

Write the letter beside each question above its answer in the box.

I   26 mm = **2·6** cm   **W**   450 cm = ____ m   **O**   400 mm = ____ m   **Y**   2340 g = ____ kg

M   80 mℓ = ____ ℓ **A**   8000 kg = ____ t   **N**   4·5 t = ____ kg   **F**   2·34 t = ____ kg

L   72 kg = ____ t   **R**   8 ha = ____ m^2 **E**   7·2 ha = ____ m^2 **S**   8234 m^2 = ____ ha

H   8 ℓ = ____ cm^3 **B**   80 mℓ = ____ cm^3 **T**   0·26 m^3 = ____ ℓ **P**   45 cm^3 = ____ ℓ

B *Timely teasers*

a   Change

  i   292 months to years and months _____

  ii   1153 minutes to hours and minutes _____

b   Kyra was knitting a scarf. She started on 1 January 2004 and knitted 1 cm per day.
  What was the date when the scarf was

  i   35 cm long _____   **ii**   71 cm long? _____

2004 was a leap year.

C *Wordy wonderings*   Remember the units.

a   A farm is 800 m by 450 m. How many hectares is this? _____

b   Malcolm pays for 40 ℓ of petrol. He puts 2500 cm^3 in his lawn mower, and the rest in his car.
  How many cm^3 of petrol does he put in his car? _____

c   A ferry weighs 96 tonnes. 40 people, weighing an average of 70 kg each, board the ferry.
  What is the mass of the loaded ferry in

  i   kg _____   **ii**   tonnes? _____

*d   The capacity of a shipping container is 16 000 ℓ. Sacks of wheat are poured into the container.
  If each sack holds 0·5 m^3, how many sacks will fit in the container? _____

81 Metric and Imperial Equivalents

Let's look at ...
● making approximate conversions from metric units to imperial units and from imperial units to metric units

✓ You need to know these rough **metric equivalents** and rough **imperial equivalents**.

length	mass	capacity
1 mile ≃ 1·6 km	1 pound ≃ $\frac{1}{2}$ kg	1 pint ≃ 600 mℓ
1 yard ≃ 1 m	1 oz ≃ 30 g	1 gallon ≃ 4·5 ℓ
1 inch ≃ 2·5 cm		

length	mass	capacity
1 km ≃ 0·625 miles	1 kg ≃ 2·2 lb	1 ℓ ≃ 1·75 pints
(8 km ≃ 5 miles)		
1 m ≃ 1 yard or 3 feet		

Remember all of these are approximate.

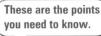

These are the points you need to know.

A Pick the best

Only use a calculator if you need to.

For each measurement decide whether Bella, Carl or Dominic has made the best conversion.

	Measurement	Bella's conversion	Carl's conversion	Dominic's conversion	Best conversion
a	5 oz	30 g	35 g	150 g	
b	10 inches	2·5 cm	4 cm	25 cm	
c	12 pounds	6 kg	12 kg	24 kg	
d	4 gallons	18 ℓ	1 ℓ	45 ℓ	
e	5 kg	2 lb	11 lb	22 lb	
f	24 km	10 miles	38 miles	15 miles	
g	9 m	3 feet	27 feet	3 yards	
h	60 pints	10 mℓ	3600 mℓ	36 ℓ	

B Convert it

a Merle uses lengths of ribbon and lace when she sews. Find the approximate metric equivalent of these lengths.

   i 4 inches _____   **ii** 20 inches _____   **iii** 6 feet _____   **iv** 15 feet _____

b Babies born at a hospital are weighed in kilograms. Find the approximate equivalents in pounds.

   i 4 kg _____   **ii** 3 kg _____   **iii** 2·5 kg _____   **iv** 4·5 kg _____

c Buckets have the following capacities. Find the approximate metric equivalents of these.

   i 2 gallons _____   **ii** 2 pints _____   **iii** 20 pints _____   **iv** 2·5 gallons _____

C Order It

a Four friends have a competition to see who can drive a golf ball the furthest.
Decide who came first, second, third and fourth.
   Nathan: 240 m   Kishan: 600 feet   John: 248 m   Antony: 660 feet
First _____   Second _____   Third _____   Fourth _____

b Harriet lives in Salisbury. She consults some maps to see how far away four neighbouring towns are.
   Deptford: 19 km   Ringwood: 26 km   Stockbridge: 15 miles   Upavon: 17 miles
Write the names of the towns in order from nearest to furthest away.

_____

How did you find this?   EASY   OK   HARD

82 Units, Measuring Instruments and Accuracy

Let's look at ...
● choosing the degree of accuracy when measuring

These are the points you need to know.

✓ When measuring we must choose the **degree of accuracy**, the **unit** and a suitable **measuring instrument**.

Example When measuring the length of a pin we could measure it to the nearest mm using a ruler.

✓ The **degree of accuracy** chosen depends on the situation.

Example A jeweller might need to measure to the nearest mm.
A surveyor might need to measure to the nearest m.

A Match it up

Choose a sensible degree of accuracy for each of these. Choose from the box.

a Mass of an armchair **C**_____

b Distance from London to Bath _____

c Capacity of a spoon _____

d Width of a pencil lead _____

e Time to swim 100 m _____

f Mass of a diamond ring _____

g Mass of a large truck _____

h Height of a table _____

i Length of a playing field _____

j Mass of a roast chicken _____

k Capacity of a bath _____

l Length of a television programme _____

A	nearest g
B	nearest 100 g
C	nearest kg
D	nearest tonne
E	nearest ml
F	nearest 100 ml
G	nearest l
H	nearest mm
I	nearest cm
J	nearest m
K	nearest 100 m
L	nearest km
M	nearest second
N	nearest minute
O	nearest 5 minutes
P	nearest hour

B Tanya's family weigh in

Do you think these measurements are given to

A the nearest g   **B** the nearest 10 g   **C** the nearest $\frac{1}{2}$ kg

D the nearest kg   **E** the nearest 10 kg   **F** the nearest 100 kg?

a Tanya's pet beetle weighs 6 g. _____

b Tanya's big brother weighs 70·5 kg. _____

c Tanya's car weighs 1200 kg. _____

* C Planet Mercury

Decide whether these measurements are given to the nearest

A km   **B** 10 km   **C** 100 km   **D** 1000 km   **E** 1 000 000 km.

a The diameter of Mercury is 4878 km. _____

b Caloris Basin, on Mercury, is 1300 km across. _____

c On average, Mercury is 58 000 000 km from the sun. _____

How did you find this?   EASY   OK   HARD

83 Estimating

Let's look at ...
● using benchmarks to estimate a measurement

These are the points you need to know.

✓ When **estimating a measurement** we often use benchmarks (known measurements) to compare against.

Example Juliet said, 'The approximate height of a door is 2 m. My chair is a little less than half the height of a door. I estimate the height of my chair to be between 80 cm and 1 m.'

Ⓐ *Explain yourself*

● The height of a door is about 2 m.
● The capacity of a teaspoon is about 5 mℓ.
● The capacity of a glass is about 250 mℓ.
● The mass of a larger bag of sugar is about 1 kg.
● The mass of a small car is about 1000 kg (or 1 tonne).
● The area of a piece of A4 paper is about 600 cm^2.

Give a range for each estimate.

Estimate each of the following by comparing it with one of the benchmarks in the above box.
Explain how you made your estimate.

a the capacity of a soup spoon _____

_____

b the height of the ceiling in your bedroom _____

_____

c the mass of a loaf of bread _____

_____

d the capacity of a jam jar _____

_____

e the mass of a bus _____

_____

f the area of a postcard _____

_____

Ⓑ *Quick questions*

Using metric units, write down an approximate measurement for each of these.
Give a range for each.

a width of a door _____

b height of a bed _____

c capacity of a dessert bowl _____

d mass of this book _____

e mass of a CD _____

f capacity of a baby bath _____

g time to run 400 m _____

h time to walk 1 km _____

i area of a pillow case _____

How did you find this?   EASY   OK   HARD

84 Bearings

Let's look at ...
● using bearings to specify direction

✓ **Bearings** are always taken in a clockwise direction from North and always have three digits.

Example

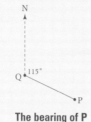

The bearing of B
from A is 052°.

The bearing of P
from Q is 115°.

The bearing of Q from P
is 180° + 115° = 295°.

Because alternate
angles are equal.

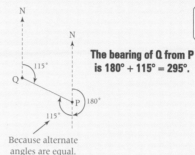

These are the points
you need to know.

(A) ***Compass talk***

This diagram shows the points of the compass.

The angle between any two adjacent compass points is 45°.

a The bearing 090° is the same as   *East*   .

b The bearing 180° is the same as   _____ .

c The bearing 045° is the same as   _____ .

d West is the same as a bearing of   _____ .

e South east is the same as a bearing of   _____ .

(B) ***Get your bearings***

Write down the bearing of P from Q **and** the bearing of Q from P for each of these.

a

b

c

d

i P from Q = ____   **i** P from Q = ____   **i** P from Q = ____   **i** P from Q = ____

ii Q from P = ____   **ii** Q from P = ____   **ii** Q from P = ____   **ii** Q from P = ____

(C) ***Island adventures***

Use your protractor to measure the bearings of these.

a Crystal Island from Shark Bay. _____

b Pirate's Cove from Crystal Island. _____

c Shark Bay from Crystal Island. _____

d Crystal Island from Pirate's Cove. _____

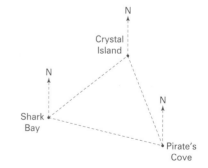

---

*(D) ***Lily's bike ride***

Lily cycles 4 km from home to the shop on a bearing of 055°.

She then cycles 5 km from the shop to
the park on a bearing of 105°.

a Draw Lily's bike ride starting at her
home, using a scale of 1 cm = 1 km.

b What is the bearing of the park from
Lily's home? _____

Lily's
home

How did you find this?   **EASY**   **OK**   **HARD**

85 Area and Perimeter

Let's look at ...
● calculating the area and the perimeter of triangles, parallelograms and trapeziums

These are the points you need to know.

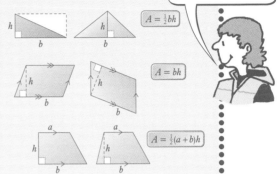

✓ **Area of a triangle** = $\frac{1}{2}$ area of rectangle
  = $\frac{1}{2}$ × base × height
  = $\frac{1}{2}bh$

The height and base must be perpendicular.

$A = \frac{1}{2}bh$

✓ **Area of a parallelogram** = bh where b is the base of the parallelogram and h is the perpendicular height.

$A = bh$

✓ **Area of a trapezium** = $\frac{1}{2}(a+b)h$
  where a and b are the parallel sides of the trapezium and h is the perpendicular height.

$A = \frac{1}{2}(a+b)h$

A Triangles

 Remember the units.

Calculate the area (A) of each triangle. Calculate the perimeter (P) of the triangles in **a** and **c**.

a

15 m 12 m
9 m

A = _____ m^2
P = _____ m

b

12 cm
20 cm

A = _____

c

7 m 9·9 m

A = _____
P = _____

d

5 mm 8 mm

A = _____

e

5 cm

A = _____

B The 'Crazy House'

In a fun park, the 'Crazy House' is built with windows having a variety of shapes.
Calculate the missing numbers.

a
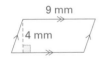
9 mm
4 mm

Area = _____

b

3·6 cm 2·5 cm

Area = _____

c
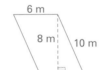
6 m
8 m 10 m

Area = _____
Perimeter = _____

d

h
6 cm

Area = 42 cm^2
h = _____

e

4 m
5 m
8 m

Area = _____

f
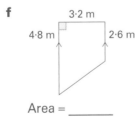
3·2 m
4·8 m 2·6 m

Area = _____

g

5·2 m
1·4 m   0·7 m
4·9 m

Area = _____
Perimeter = _____

*C Puzzles

a Calculate the area of quadrilateral Q.

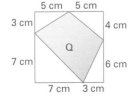
5 cm 5 cm
3 cm   4 cm
Q
7 cm   6 cm
7 cm 3 cm

 HINT: Calculate the areas of the four triangles.

Area of Q = _____

b The area of a triangle is 240 mm^2.
What might the values of b and h be?

 Try to find at least five possible answers.

1 b = _____ h = _____   **2** b = _____ h = _____
3 b = _____ h = _____   **4** b = _____ h = _____
5 b = _____ h = _____

How did you find this? EASY OK HARD

86 Volume and Surface Area

Let's look at ...
- calculating the volume and surface area of cuboids
- calculating the volume of shapes made from cuboids

✓ Remember, the **surface area** of a solid is the total area of all the faces.
The **surface area of a cuboid** = 2 (length × width) + 2(length × height) + 2(height × width)
= **2lw + 2lh + 2hw**.

✓ **Volume** is the amount of space taken up by something.
Volume is measured in mm³, cm³, m³ or ℓ, pints, gallons.
Volume of a cuboid = length × width × height
= **lwh**

These are the points you need to know.

A Pack it

a Find the volume (V) and the surface area (A) of the box, the briefcase and the suitcase.

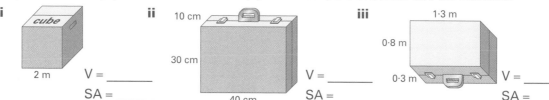

i

cube

2 m

V = _____
SA = _____

ii

10 cm

30 cm

40 cm

V = _____
SA = _____

iii

1·3 m

0·8 m

0·3 m

V = _____
SA = _____

b Madame Dupont is packing pocket dictionaries into a box as shown.

i How many dictionaries will fit on the bottom layer of the box? _____

ii How many dictionaries will fit in the box altogether? _____

iii What is the surface area of the box? _____
Show your working.

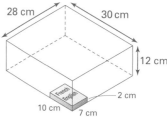

28 cm   30 cm

12 cm

French English

2 cm

10 cm   7 cm

B Lucy Ella Hamilton

Lucy made her initials out of wood. Find the volume of each letter.

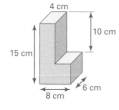

4 cm

10 cm

15 cm

8 cm   6 cm

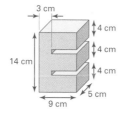

3 cm

4 cm

4 cm

4 cm

14 cm

9 cm   5 cm

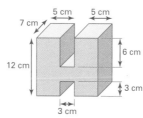

5 cm   5 cm

7 cm

6 cm

12 cm

3 cm

3 cm

Volume of

L = _____

E = _____

H = _____

C Puzzle

a What is the total surface area of this box? _____

b Find the length of each side. _____, _____, _____

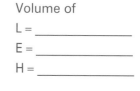

120 m²

60 m²

18 m²

D Your turn

Find two cuboid shaped boxes. Sketch each box and label the length of the sides.
Calculate the volume and surface area of each box.

How did you find this?   EASY   OK   HARD

89

HANDLING DATA

87 Discrete and Continuous Data

Let's look at ...
- the difference between discrete and continuous data
- grouping continuous data on a frequency table

These are the points you need to know.

✓ **Discrete** data can only have certain values.

   Example   The number of crisps in crisp packets.

✓ **Continuous** data can have any values within a certain range.

   Example   The diameter of crisps at their widest part.

✓ When we collect continuous data we usually group it into **equal class intervals** on a **frequency chart**.

Speed of cars passing the school

Speed (s) mph	Tally	Frequency
10 < s ⩽ 15	⊞⊞ ⊞⊞ III	13
15 < s ⩽ 20	⊞⊞ ⊞⊞ ⊞⊞ II	17
20 < s ⩽ 25	⊞⊞ II	7
25 < s ⩽ 30	IIII	4
30 < s ⩽ 35	I	2

A *Discrete or continuous?*

Decide whether each of the following are discrete data (D) or continuous data (C).

a the height of sunflowers _____

b the number of pupils with brown eyes _____

c the mass of rocks in a collection _____

d the amount of water drunk in a day _____

e the population of towns in Cornwall _____

f the time spent doing homework _____

g the length of earthworms _____

h the number of people at concerts _____

i the cost of hamburgers _____

j the areas of lakes in Scotland _____

B *Weekly groceries*

Jacob timed customers at his local supermarket. He wrote down the number of minutes they spent in the supermarket, to the nearest minute.

17	9	48	32	8	3
43	52	50	6	38	20
13	5	30	37	44	41

Time in supermarket (mins)	Tally	Frequency
0 ⩽ t < 10		
10 ⩽ t < 20		
20 ⩽ t < 30		
30 ⩽ t < 40		
40 ⩽ t < 50		
50 ⩽ t < 60		

a Complete the frequency chart.

b How many customers had a time in the class interval 40 ⩽ t < 50? _____

c How many customers spent less than 20 minutes in the supermarket? _____

d How many customers spent at least 30 minutes in the supermarket? _____

C *Catch a cab*

Susie recorded the number of kilometres she drove in her taxi each day.

24·3	161·3	98·2	49·2	158·9	50
210·8	192·5	82	99·7	108·1	123·4
182·3	142·7	150·1	200	0	134·6

Distance travelled (km)	Tally	Frequency
0 ⩽ t < 50		
50 ⩽ t < 100		

a Complete the frequency chart.

b How many days did Susie drive greater than or equal to 100 km but less than 150 km? _____

c How many days did Susie drive less than 200 km? _____

d Explain why the class intervals on the table are more useful than the intervals 0 ⩽ t < 10, 10 ⩽ t < 20, 20 ⩽ t < 30, ... _____

88 Two-way Tables

Let's look at ...
● interpreting and designing two-way tables

These are the points you need to know.

✓ A **two-way table** displays two data sets in a table.

Example This two-way table shows the ages of students in school sports teams.

	Netball	Football	Hockey
Under 14	16	25	27
14 to 16	36	53	26
Over 16	29	34	28

A *Learning languages*

Pupils in Year 8 at St. Thomas' College chose which language they wanted to study. This table shows their choices.

	Boys	Girls
French	8	12
German	21	14
Latin	5	9

a How many girls chose French? _____

b How many pupils chose Latin? _____

c How many boys were in Year 8? _____

B *Summer sports*

Millgate School takes part in a summer sports tournament every year. This table shows their results in the three sports.

	Cricket	Tennis	Volleyball
Win	9	2	4
Draw	6	1	5
Lose	2	7	5

a How many volleyball games did they win? _____

b How many games did they draw altogether? _____

c In which sport did Millgate School do the best? _____

d How many games were played in total? _____

C *All day café*

Cost of meal	Breakfast	Lunch	Dinner
£0–£9·99	9	4	0
£10–£19·99	3	10	4
£20–£29·99	0	1	13
£30–£30·99	0	0	3

This table shows the costs of meals (including drinks) at the 'Strawberry Café' one day.

a How many lunches cost less than £10? _____

b How many meals cost greater than or equal to £20 but less than £30? _____

c How many lunches cost less than £30? _____

d How many people dined at the 'Strawberry Café' in total on this day? _____

D *'The Lonely Knight'*

Felicity's school put on two performances of 'The Lonely Knight', one on Friday and one on Saturday. After the show people were asked to rate it excellent, good or poor.
Design a two-way table for Felicity to collect people's ratings after each of the performances.

How did you find this? EASY OK HARD

89 Surveys – Collecting the Data

Let's look at ...
● answering a question by exploring related questions
● methods to collect data

✓ When solving a problem using statistical methods, think of other **related questions**. This helps you decide what data needs to be collected.

✓ You can **gather data** by
● doing a survey using a questionnaire or collection sheet
● doing an experiment where you count or measure something
● gathering data from other sources such as books, newspapers, the Internet ...

These are **primary sources**.

These are the points you need to know.

✓ The **sample size** should be as large as is sensible. If it is too small it is not representative. How big it can be depends on time, money and who the survey is for.

These are **secondary sources**.

A Planning charts

Complete a planning chart for each of these two questions.

1 How much sport do children play?	**2 Are ancient or modern tourist attractions more popular?**
a Other related questions	**a** Other related questions
b Some possible results	**b** Some possible results
c Data to be collected	**c** Data to be collected
d Collection sheet or questionnaire *Design yours and draw it here.*	**d** Collection sheet or questionnaire
e Suitable sample size	**e** Suitable sample size
f Is the data source primary or secondary?	**f** Is the data source primary or secondary?

90 Mode and Range

Let's look at ...
● calculating the mode and the range
  for discrete and continuous data

✓ The **mode** is the most commonly occurring data value.
  If a frequency chart has grouped data we find the **modal class**.

Example

Age	0–	5–	10–	15–	20–
Frequency	2	3	19	26	21

The modal class is 15– because this class interval has the highest frequency.

✓ The **range** is the difference between the highest and lowest data values.
  It is a simple measure of the **spread** of the data.
  The **range** of a set of continuous data is the **highest rounded value minus the lowest rounded value**.

Example   The highest age, rounded to the nearest year, is 23 years.
          The lowest age, rounded to the nearest year, is 2 years.
          The range is 23 – 2 = 21 years.

These are the points you need to know.

A Cats and dogs and ...

This chart gives the number of pets owned by pupils in a class.

a What is the modal number of pets owned? _____

b What is the range of the number of pets owned? _____

Number of pets	Frequency
0	8
1	11
2	6
3	3
4	1

B Fitness test

Miss West asked her P.E. class to do a fitness test in February, and again in May.
Each pupil had to run as far as they could in 8 minutes. This chart shows how far the pupils ran in metres.

Distance (d)	February Frequency	May Frequency
1200 ⩽ d < 1400	2	1
1400 ⩽ d < 1600	6	4
1600 ⩽ d < 1800	8	5
1800 ⩽ d < 2000	4	10
2000 ⩽ d < 2200	3	1
2200 ⩽ d < 2400	2	4

a What was the modal class for
  i February? _____
  ii May? _____

b In February the shortest distance run was 1212 m and the longest distance was 2203 m.
  What was the range? _____

c In May the shortest distance run was 1293·7 m and the longest distance was 2393·4 m.
  What was the range? _____

d Why would Miss West want to calculate the modes and the ranges? _____

C Big apples

Two varieties of apples are grown by an orchard. Clare selects ten apples of each variety and weighs them. The masses are shown in this chart.

Masses of variety A(g)	153·6	161·2	184·3	175·9	223·6	141·6	229·2	192·9
Masses of variety B(g)	177·3	184·7	179·6	187·2	190·1	201·3	194·9	198·3

a Find the range for   **i** variety A _____   **ii** variety B _____

b Use the ranges to compare the two varieties _____
_____

How did you find this?   EASY   OK   HARD

91 Mean

Let's look at ...
- finding the mean from a frequency table
- calculating the mean using the assumed mean method

✓ The **mean** gives an idea of what would happen if there were equal shares.

$$\text{mean} = \frac{\text{sum of data values}}{\text{number of values}}$$

Sometimes the data is given in a frequency table.

Score	0	1	2	3	4
Frequency	3	5	4	6	9

Example   Mean $= \dfrac{0 \times 3 + 1 \times 5 + 2 \times 4 + 3 \times 6 + 4 \times 9}{3 + 5 + 4 + 6 + 9}$

total of (score × frequency)

total frequency

$= \dfrac{67}{27} = 2 \cdot 48$ (2 d.p.)

✓ We can find the mean using an **assumed mean**.

Example   5·6, 6·2, 6·8, 5·3, 6·9, 4·8, 6·4, 6·5

These are the points you need to know.

To find the assumed mean

1 assume the mean is 6

2 subtract assumed mean from each data value   ⁻0·4, 0·2, 0·8, ⁻0·7, 0·9, ⁻1·2, 0·4, 0·5

3 find mean of differences $= \dfrac{⁻0·4 + 0·2 + 0·8 + ⁻0·7 + 0·9 + ⁻1·2 + 0·4 + 0·5}{8} = 0 \cdot 0625$

4 add your answer to part 3 to assumed mean $6 + 0 \cdot 0625 = 6 \cdot 0625$

A Car pool

Tori counted the number of people in cars which passed her school.

Number of people	1	2	3	4	5	6	7
Frequency	72	36	19	5	8	1	2
Number of people × frequency	1 × 72						

This table shows the results.
Calculate the mean number of people in cars, to 2 d.p. _____

B Show your working

Calculate the mean of each of these using the assumed mean method.
Show your working.

a 9, 11, 15, 17

b 13, 6, 5, 9, 4, 8

c 8·2, 8·6, 8·3, 8·4, 8·9

C Competition time

This table shows how long it took five boys and five girls to complete a puzzle.

a Find the mean time for the girls _____

b Find the mean time for the boys _____

Girls' times (secs)	22	17	26	21	29
Boys' times (secs)	14	31	18	28	31

c Find the mean time for all 10 pupils. _____

Could you find this by finding the mean of the answers to **a** and **b**? _____

92 Median, Mean, Mode, Range

Let's look at ...
● finding the median, mean, mode and range of real-life data

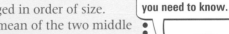

These are the points you need to know.

✓ The **median** is the middle value when a set of data is arranged in order of size. When there is an even number of values, the median is the mean of the two middle values.

Example   5, 6, 3, 9, 6, 4, 2, 1

middle values

In order these are   1, 2, 3, (4, 5), 6, 6, 9.

Median = $\frac{4+5}{2}$ = 4·5

✓ The **mean** gives us an idea of what would happen if there were 'equal' shares.

✓ The **mode** is useful for identifying the 'most popular'.

✓ The **range** is a measure of the spread of the data.

A Long and short

Tom and his friends each measured the length of their right foot. This list gives the lengths in mm.

157   163   158   142   193   202   170   175   161   188   147   179

a Find the median length. _____

b Complete this sentence.

About half of the friends' right feet are longer than _____ mm.

c Find the mean length to 2 d.p. _____

d Find the range of the lengths. _____

B Early and late

A police radar camera was set up to record the speed of cars travelling down High St at 8:00 a.m. and again at 8:00 p.m. This table shows the speeds.

| Speeds at 8:00 a.m. (miles/hour) | 14 | 11 | 18 | 9 | 13 | 8 | 14 | 21 | 17 | 16 |
| Speeds at 8:00 p.m. (miles/hour) | 31 | 25 | 34 | 29 | 33 | 26 | 22 | 34 | 28 | 41 |

a Find the mean, median and range for each time.

   i   8:00 a.m   Mean _____   Median _____   Range _____

   ii 8:00 p.m   Mean _____   Median _____   Range _____

b On average, did the cars on High St travel faster in the morning or in the evening? _____

Why do you think this might be? _____

_____

C Puzzle

Ezra has five number cards. Four of the number cards are shown. What number is on the fifth card if

a the mode is 2? _____

b the range is 8? _____

c the mean is 5? _____

d the mean is the same as the median? _____

b and d have 2 possible answers. Find them both.

How did you find this?   EASY   OK   HARD

93 Finding the Median, Range and Mode from a Stem-and-leaf Diagram

Let's look at ...
● constructing and using stem-and-leaf graphs

These are the points you need to know.

✔ We can find the median, range and mode from a **stem-and-leaf graph**.

Example There are 25 data values.
The median is the 13th value, 0·8 hours.
The range is 4·0 − 0·2 = 3·8 hours.
The mode is 0·5 hours.

Time taken for lunch
stem = hours, leaves = tenths

```
0 | 2 2 4 5 5 5 5 5 5 6 6 7 7 8
1 | 0 0 0 0 5 5 6
2 | 0 5
3 | 1 5
4 | 0
```

stem   leaves

The 13th value has a stem of 0 and a leaf of 8. It is 0·8 hours.

A Getting on

This stem-and-leaf graph shows the ages of the 17 members of Mark's football team.

a Find the median age. _____

b Find the range of ages. _____

c Find the modal age. _____

Ages of football players (years)

```
1 | 8 8 9
2 | 0 1 1 1 3 8 9
3 | 0 3 3 4 6        stem = tens
4 | 2 3              leaves = units
```

B Fastest finger first

Contestants for a television quiz show had to complete a timed multi-choice test. The times taken to complete the test are shown in this stem-and-leaf graph.

a Find the median time taken. _____

b Find the range of the times. _____

c Find the modal time taken. _____

Times to complete test (secs)

```
0 | 9 9
1 | 0 1 3 3 4 8 9 9
2 | 0 0 2 2 5 7 7 7 8 9
3 | 0 1 1 2 6 8 9
4 | 2 3 3 4 5
5 | 0 1 7          stem = tens
                   leaves = units
```

C Farmer Brown's hens

Natasha Brown recorded the masses of eggs laid by her 33 hens one week. This list shows the masses in grams.

53	55	61	73	85	92	58	63	80	73	62	94	87	88	75	64	74
67	72	60	53	81	90	89	79	90	61	54	90	93	85	83	86	

a Complete the stem-and-leaf diagram for this data.

b How many eggs weighed from 80 to 89 grams? _____

c How many eggs weighed less than 70 grams? _____

d Find the median egg mass. _____

e Find the range of egg masses. _____

f Find the modal egg mass. _____

Masses of Farmer Brown's eggs (grams)

```
5 | 3
6 |
7 |
8 |          stem = ____
9 |          leaves = ____
```

94 Comparing Data

Let's look at ...
● using the range and the mean, median or mode to compare data

✓ To **compare data** we use the range and one or more of the mean, median or mode.

Example Robert found the mean, median, mode and range of the hours of sunshine per week in two places

Place A mean 53 range 24 **Place B** mean 54 range 8

The two places have about the same mean sunshine hours but place A has a much less consistent number of sunshine hours.

A Which way?

Oliver and his family can either take an M road or an A road to visit Oliver's Nana.

	Mean time (mins)	Median time (mins)	Range (mins)
M road	54	54	16
A road	56	55	4

Which road would you suggest they use? You can choose either as long as you use the results in the table to explain why. _____

_____

_____

B Ten scoops please

Marcia and her brother Fred disagree about which local Fish and Chip shop sells the best-value-for money chips.
Both shops charge the same amount for one scoop.
Marcia and Fred ordered 10 scoops of chips from each shop and weighed them.
This table lists the results.

	Mean of scoop (g)									
Neptune's	695	627	593	581	634	575	678	601	659	657
Cousteau's	641	652	648	622	629	632	663	649	627	637

a Calculate the mean and the range for each shop.

Neptune's _____

Cousteau's _____

b Which shop do you think is the best value for money? Explain why using your results from **a**.

_____

_____

C Jump off

Class 8E hold a vote to decide which girl will represent them in an inter-class long jump competition.
The top two jumpers are Nellie and Leah.
The results of their last five jumps are listed below.

| Nellie | 2·2 m | 2·2 m | 2·15 m | 2·2 m | 2·25 m |
| **Leah** | 2·3 m | 2·35 m | 2·35 m | 1·1 m | 2·4 m |

	Mean	Median	Mode	Range
Nellie				
Leah				

a Complete the table.

b Which girl would you vote for? Use the table in **a** to explain why. _____

_____

_____

How did you find this? **EASY** **OK** **HARD**

95 Compound Bar Charts and Line Graphs

Let's look at ...
● drawing and understanding compound bar charts and line graphs

✓ A **compound bar chart** is used to show categorical data.

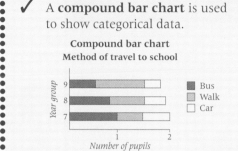

Compound bar chart
Method of travel to school

- Bus
- Walk
- Car

Year group
Number of pupils (in hundreds)

✓ A **line graph** is used to show changes over time.

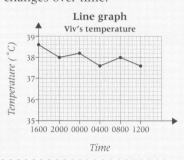

Line graph
Viv's temperature

Temperature (°C)
Time

These are the points you need to know.

A *Travel times*

This table shows the usual time taken to travel to work by full-time and part-time adults working away from home.

	Percentages	
Travel time	**Full-time**	**Part-time**
10 minutes or less	24	39
11–20 minutes	24	25
21–30 minutes	16	11
31–40 minutes	5	2
41–50 minutes	6	3
51–60 minutes	5	2
more than one hour	4	1

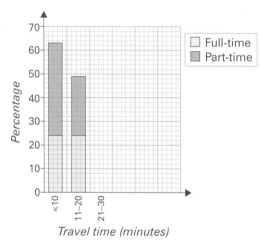

- Full-time
- Part-time

Percentage
Travel time (minutes)

a Complete this compound bar chart for the above data.

b Give the graph a title.

c Overall do full-time or part-time workers travel for longer to get to work? _____

B *Call an ambulance!*

This table gives the number of emergency calls made to the ambulance service in England, and the number of calls that the ambulance service actually responded to.

Year	Calls made (millions)	Responses (millions)
1995/96	3·2	2·9
1996/97	3·3	3·0
1997/98	3·6	3·2
1998/99	3·8	3·3
1999/2000	4·2	3·4
2000/01	4·4	3·6
2001/02	4·7	3·8

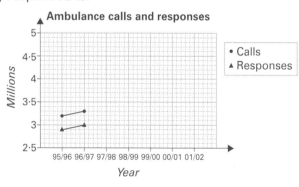

Ambulance calls and responses

- Calls
- Responses

Millions
Year

a Complete this line graph with both sets of data on the same grid.

b Comment on the trends shown by the graph. _____

_____

How did you find this? [EASY] [OK] [HARD]

96 Frequency Diagrams

Let's look at ...
● drawing and interpreting frequency diagrams

✓ For **continuous** data we draw a **frequency diagram**.
We label the divisions *between* the bars.

Remember continuous data can take any value within a given range.

Example Harriet noticed that teachers at her school were often late back to class after lunch. This table shows how many minutes late teachers at her school were one day.

Minutes late (m)	Frequency
$0 \leqslant m < 2$	6
$2 \leqslant m < 4$	4
$4 \leqslant m < 6$	5
$6 \leqslant m < 8$	2

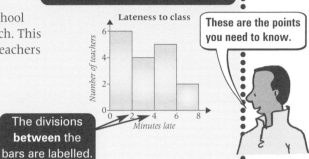

These are the points you need to know.

The divisions **between** the bars are labelled.

A How big is a glass?

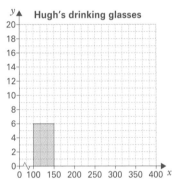

Hugh noticed that his family had quite an assortment of drinking glasses. He measured the capacity of each glass.

Capacity (mℓ)	Number of glasses
$100 \leqslant c < 150$	6
$150 \leqslant c < 200$	5
$200 \leqslant c < 250$	18
$250 \leqslant c < 300$	12
$300 \leqslant c < 350$	7
$350 \leqslant c < 400$	4

a Complete the frequency diagram. Remember to label the axes.

b How many of the glasses had a capacity of at least 350 mℓ? _____

c How many of the glasses had a capacity of less than 200 mℓ? _____

d How many glasses were there altogether? _____

B Gooseland

The head of marketing at Gooseland Fun Park wanted to know how many hours people travelled before arriving at Gooseland. She surveyed people as they arrived.

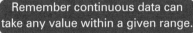

0– means $0 \leqslant t < 0.5$ and 0.5– means $0.5 \leqslant t < 1$ ...

Time travelled (hours)	0–	0·5–	1–	1·5–	2–	2·5–3
Frequency	140	174	68	142	36	12

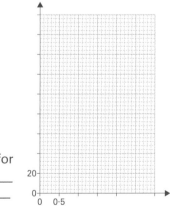

a Draw a frequency diagram for this data.
Remember to give the graph a title and label the axes.

b How many people travelled for less than 1 hour? _____

c How many people travelled for at least 2 hours? _____

d The head of marketing wanted to know how many people travelled for less than $\frac{3}{4}$ of an hour. Can she tell this from the graph? _____
Why or why not? _____

e What was the modal class for time travelled? _____

f Give a reason why the head of marketing might want to know the modal class for time travelled.

_____

How did you find this? EASY OK HARD

97 Drawing Pie Charts

Let's look at ...
● displaying categorical data on a pie chart

✓ A **pie chart** is a graph represented by a circle. The bigger the sector of the circle the bigger the proportion it represents.

✓ To **draw a pie chart**
● find what fraction of the whole each sector should represent
● multiply this fraction by 360° to find the angle at the centre of the sector.

✓ Pie charts are mostly suitable for **categorical (non-numerical) data**.

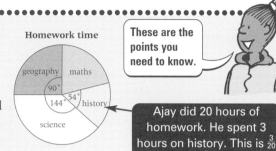

Homework time

These are the points you need to know.

Ajay did 20 hours of homework. He spent 3 hours on history. This is $\frac{3}{20}$ of the time. $\frac{3}{20} \times 360° = 54°$

(A) Categorical data

Which **two** of these sets of data would be most suitable to display on a pie chart?
A number of videos watched in a week.
B proportion of students with different coloured eyes.
C proportion of hours different brands of batteries last.
D proportion of households reading different newspapers.   ☐ and ☐

(B) Fruity gums

This table shows the number of different flavours of fruity gums in Marietta's packet.

Flavour	Orange	Lemon	Strawberry	Lime	Total
Number	12	9	4	5	30

Marietta wanted to draw a pie chart.

a Fill in the missing numbers to calculate the angles of the sectors.
  i orange
    $\frac{12}{30} \times 360° = 144°$
  ii lemon
    $\frac{9}{30} \times 360° = \boxed{}$
  iii strawberry
    $\frac{\boxed{}}{30} \times \boxed{} = \boxed{}$
  iv lime
    $\frac{\boxed{}}{\boxed{}} \times \boxed{} = \boxed{}$

b Show the fruity gum flavours on a pie chart.

Remember to label each sector.

(C) CD collection

Kieren sorted his CD collection into five different styles of music. This table shows how many CDs of each style he had.

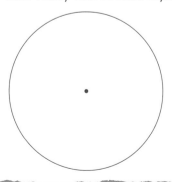

Music style	Number of CDs	Pie chart angle
Jazz	8	
Rock	17	
Classical	5	
Hip Hop	11	
Country	4	
Total	45	360°

a Calculate, then fill in the pie chart angles on the table.

b Represent the information on a pie chart.

98 Scatter Graphs

Let's look at ...
● drawing and interpreting scatter graphs

✓ A **scatter graph** displays two sets of data.

Example This scatter graph shows the mass of people and the mass of pizza each ate at a pizza night.
It shows that heavier people tend to eat more pizza.

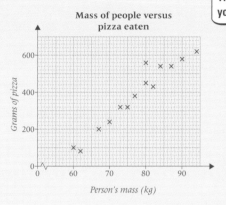

These are the points you need to know.

A After school

Lydia asked ten students for the time they spent watching television and the time they spent doing homework each week.

| Time watching TV (hours) | 24 | 8 | 2 | 20 | 12 | 26 | 9 | 18 | 22 | 16 |
| Time doing homework (hours) | 9 | 15 | 16 | 6 | 10 | 2 | 13 | 9 | 5 | 11 |

a Use the information in the table to draw a scatter graph.

b

Lydia said, *'As a general trend, the longer you spend watching television, the more homework you do.'*

Is Lydia correct or incorrect? _____ Explain why. _____
_____

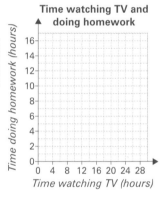

B Hit the brakes!

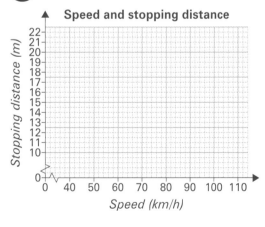

Constable Black recorded the speeds and stopping distances of eight cars.
The stopping distance is the shortest distance it takes for the car to stop after the driver has put his or her foot on the brakes.

| Speed (km/hr) | 50 | 102 | 68 | 84 | 92 | 42 | 110 | 72 |
| Stopping distance (m) | 10 | 19 | 11 | 15 | 16 | 8 | 21 | 14 |

a Draw a scatter graph for this data.

b What happens to the stopping distance as the speed increases? _____
_____

How did you find this? [EASY] [OK] [HARD]

99 Interpreting Graphs

Let's look at ...
● using a graph to help interpret and compare data

These are the points you need to know.

✓ We often use graphs to help us **interpret** data.

Example Robyn found these pie charts in a book.
They show the proportion of foot, car, taxi and cycle traffic for two cities of similar population.

She used the graphs to make some comparisons.

There was about twice as much foot traffic in city 1.
There was about half as much car traffic in city 1.

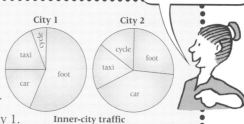

Inner-city traffic

Ⓐ *Snack time*

These pie charts show some nutritional information for two different snack foods.

a If William wants a low fat snack, which of the two treats should he choose? _____

b Describe the differences between the two snack foods.

_____
_____
_____
_____

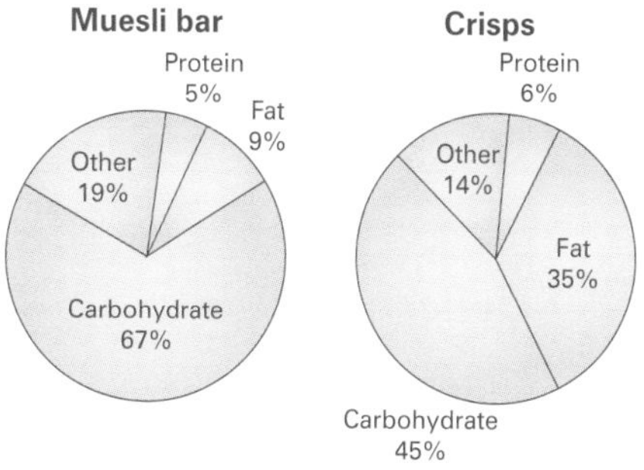

Muesli bar
Protein 5%
Fat 9%
Other 19%
Carbohydrate 67%

Crisps
Protein 6%
Other 14%
Fat 35%
Carbohydrate 45%

Ⓑ *Homework help*

This compound bar chart shows who key stage 2 and key stage 3 pupils go to for help with their homework.

a About what percentage of key stage 3 pupils get help from
 i family _____ **ii** friends _____
 iii teachers? _____

b Describe the differences between the two groups of pupils. _____

_____
_____

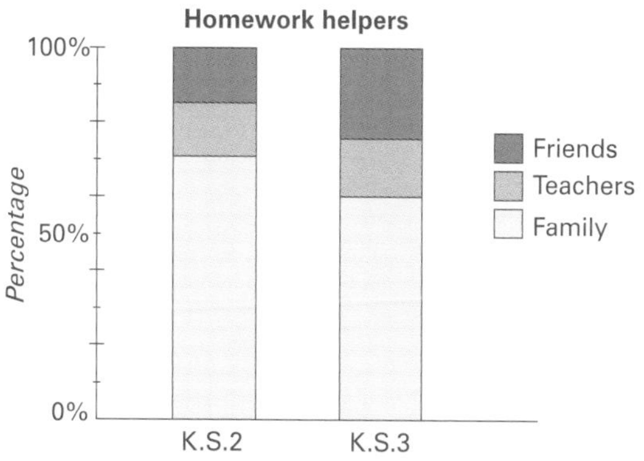

Homework helpers

Percentage

■ Friends
▨ Teachers
☐ Family

K.S.2   K.S.3

Ⓒ *Shift work*

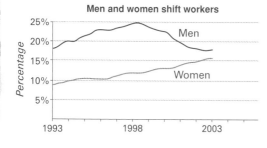

Men and women shift workers

This line graph shows the percentage of men and women in a town who do shift work.
Amy said, '*In 1993 about 9% of women did shift work. The percentage slowly increased over the next 10 years to about 16%.*'

a Describe the trend for male shift workers from 1993 to 2003.

_____
_____

b Using the graph, predict the percentage of men and women shift workers in 2005.
 Men _____   Women _____

How did you find this? [EASY] [OK] [HARD]

100 Surveying

Let's look at ...
● planning and carrying out a survey

Follow the steps below to complete your own 'mini' survey.

A Plan your survey

a What do you want to find out?

Some ideas are:
● How much sport do girls and boys play?
● How do leaf lengths on two different plants differ?
● How does the rainfall in two countries differ?

b What are some related questions?

c What data do you need to collect?

d How will you collect the data?

Will you conduct an experiment, use the internet or ask people?

e What will the sample size be?

B Collect your data

You will need extra paper for this.

Design a data collection sheet or questionnaire and collect your data. Attach it to this page.

C Display your results

Use a suitable graph if possible, or a table.

D Analyse your results

Find the mean, median, mode and range if appropriate.

E Write a summary

Write what you found out. Include your conclusions. Make sure your conclusions relate to your original question in **A**.

How did you find this?   EASY   OK   HARD

101 Language of Probability

Let's look at ...
- events which are equally likely
- outcomes which are more likely than others

These are the points you need to know.

✓ When an event is **random**, the outcome is **unpredictable**.

✓ Some outcomes are **more likely** than others.

Example This spinner is more likely to stop on white than on purple when spun.

✓ Some events have **equally likely** outcomes.

Example This spinner is equally likely to stop on 1, 2, 3, 4 or 5.

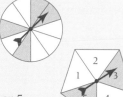

A Spin it

  A B C D

a Which spinner has an equally likely chance of stopping on light purple, dark purple or white? _____

b Which spinner is most likely to stop on **i** light purple ___ **ii** dark purple ___ **iii** white ___

B Hoop throw

A

1	2	3
4	5	6
7	8	9

B

2	3	4
5	6	7
8	9	10

C

3	4	5
6	7	8
9	10	11

At a fair you win a prize if your hoop lands on an even number.
Which grid should you throw your hoop on? _____
Explain why. _____
_____

C Card games

Box 1    **Box 2**

Box 1 and Box 2 each have eight cards in them.

a Nicola takes a card at random from Box 1. What shape is it most likely to have on it? _____

b Paul takes a card at random from Box 2. What shape is he most likely to get? _____

c Ben takes all of the cards from Box 1 and Box 2 and puts them in a bag. He takes one card from the bag without looking. What shape is he most likely to get? _____

* D Banana blues

Two bags of sweets contain different numbers of banana and orange sweets.
   Bag 1 has 13 banana sweets and 11 orange sweets.
   Bag 2 has 15 banana sweets and 9 orange sweets.
Emma does **not** like banana sweets. She can pick a sweet at random from either bag.
Which bag should she pick from? _____ Why? _____
_____

102 Calculating Probability

Let's look at ...
- probabilities of events occurring or not occurring

These are the points you need to know.

✓ Remember: If the outcomes of an event are **equally likely**, then

$$\text{probability of an event} = \frac{\text{number of favourable outcomes}}{\text{number of possible outcomes}}$$

Example   A box contains 100 counters. Seven of these are green. Ella chooses one counter at random. The probability of this counter being green is $\frac{7}{100}$ or 0·07 or 7%.

✓ If the probability of an event occurring is p, then the **probability** of it **not** occurring is $1 - p$.

Example   If the probability of having to stop at an intersection is $\frac{1}{4}$, then the probability of not having to stop is $\frac{3}{4}$.

Ⓐ *At the circus*

a The probability that a tight-rope walker will get across without falling is 95%. What is the probability that she will fall? _____

b The probability that the lions will roar during a show is 0·4. What is the probability of them not roaring? _____

c If the probability of the audience laughing at the clowns is $\frac{6}{7}$, what is the probability that they won't laugh? _____

Ⓑ *Party time*

a There are 20 balloons in a box. Eight are white, 3 are green, 4 are orange and 5 are purple. Liam takes a balloon without looking. Draw an arrow on the probability scale to show the probability that the balloon is

   i white   **ii** not white   **iii** orange or purple   **iv** neither orange nor purple

   ```
   +----+----+----+----+----+----+----+----+
   0                    1/2                  1
   ```

b Gail puts 9 red party hats and 1 blue party hat in a bag.
   April is going to take one party hat without looking.

   > She says, 'There are two colours, so it is just as likely that I will get a red hat or a blue hat.'

   i Explain why April is wrong. _____
   _____

   ii How many more blue hats should Gail put in the bag to make it just as likely that April will get a red hat as a blue hat? _____

Ⓒ *Spare change*

a Jamie has these four coins.
   If he chooses one at random, what is the probability that it is a 20p coin? _____

*b Daniel has four coins which total 17p. One of them is a 10p coin.
   If he chooses one at random, what is the probability that it is a 1p coin? _____

How did you find this?   **EASY**   **OK**   **HARD**

103 Calculating Probability by Listing Outcomes

Let's look at ...
- finding the set of all possible outcomes of two events
- calculating probabilities from the sample space

✓ To **calculate a probability** we often record all the possible **outcomes** using a list, diagram or table.

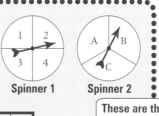

Spinner 1   Spinner 2

✓ The set of all the possible outcomes is called the **sample space**.

Example These two spinners are spun.
The possible outcomes could be shown in a table.

Spinner 1	1	1	1	2	2	2	3	3	3	4	4	4
Spinner 2	A	B	C	A	B	C	A	B	C	A	B	C

The probability of getting 1 and A = $\frac{1}{12}$ ← ways of getting 1 and A

← number of possible outcomes

These are the points you need to know.

A Summer holiday

Pete and Jenny are on holiday in Majorca.

a Pete has two pairs of shorts and four t-shirts with him.
The shorts are black and white, and the t-shirts are red, yellow, green and blue.
List all of the colour combinations Pete could wear.
black/red, _____

b Pete and Jenny can choose to swim, go shopping or play tennis.
Complete this table to show the possible outcomes.

Pete	swim	swim							
Jenny	swim	shop							

B Twins

Mrs Chen is expecting twins. She does not know the gender of the babies.

a Complete this table to show the possible outcomes.

Baby 1	girl			
Baby 2	girl			

b Find the probability of Mrs Chen having
i 2 girls _____   **ii** 2 boys _____   **iii** a boy and a girl _____

C Adding game

Rosie takes one of these cards without looking, and spins the spinner.
She adds the two numbers together.

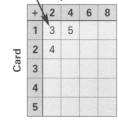

If she gets 1 on the card and 2 on the spinner they add to make 3.

	Spinner			
+	2	4	6	8
1	3	5		
2	4			
3				
4				
5				

Card

a Complete this table to show the sample space

b Find the probability that the total will be
i 4 _____   **ii** 7 _____   **iii** more than 11 _____
iv less than 9 _____   **v** at least 8 _____   **vi** even _____

104 Estimating Probability from Experiments

Let's look at ...
● using the results of an experiment to estimate probability

These are the points you need to know.

✓ We sometimes use the results of an experiment to **estimate a probability**.

Example   A factory tested five hundred circuit boards.
Twelve were found to have faults.
We can use this to estimate that the probability that a
circuit board chosen at random will be faulty is $\frac{12}{500} = \frac{3}{125}$.

Reduce fractions to their lowest terms.

Note:
1 When an experiment is repeated, there may be, and usually will be different outcomes.
2 Increasing the number of times an experiment is repeated usually leads to better estimates of probability.

(A) *Sports fanatic*

a   Brendan has won 44 out of the last 47 squash games he has played. Estimate the probability that he will win his next squash game. _____

b   Brendan has also won 35 out of the last 47 mountain bike races he has entered. Estimate the probability that he will win his next race. _____

(B) *Doubtful driving*

Write all probabilities in their simplest form.

a   A speed camera has found 50 of the last 200 passing cars to be speeding. Estimate the probability that the next car to pass will be speeding. _____

b   A police officer watching a stop sign noticed that 35 of the last 100 cars at the stop sign actually stopped. Estimate the probability that the next car will actually stop. _____

c   A busy junction recorded crashes on 90 of the last 300 days. Estimate the probability that a car crash will occur at the junction tomorrow. _____

(C) *Prize table*

Winners at a school fair can choose a prize from the prize table.
This chart shows the prizes chosen by the last 50 winners.
Estimate the probability that the next winner will choose
i   chocolates _____   **ii**   a soft toy _____

Soft toy	32
Chocolates	10
Magazine	6
Writing set	2

(D) *Full to the brim?*

Katie and Alice both did an experiment to find the proportion of milk bottles that were underfilled.

● Katie measured 10 bottles and found that 1 was underfilled.
● Alice measured 100 bottles and found that 2 were underfilled.

a   Who would have been most likely to conclude that underfilling of bottles is a problem?
_____ Explain why you think this is. _____
_____

b   Whose experiment do you think would have led to the best estimate? _____
Why do you think this? _____
_____

     How did you find this?   EASY   OK   HARD

105 Comparing Calculated Probability with Experimental Probability

Let's look at ...
- comparing the probability estimated from an experiment with the theoretical probability

✓ When we compare experimental probability with theoretical probability, the greater the number of trials in the experiment, the closer the experimental probability is to the theoretical probability.

These are the points you need to know.

A *Heads I win*

You will need a coin.

a What is the theoretical probability of getting a head when you toss your coin? _____

b Toss your coin 50 times. How many heads do you get? _____

c Is this number what you would expect? _____ Explain why or why not.

_____

d If you tossed your coin another 50 times would you expect to get the same results? _____

e What results would you expect if you tossed your coin 500 times? _____

_____

B *Red, yellow or blue*

John has 10 marbles in a bag: 5 are red, 3 are yellow and 2 are blue.
John takes a marble without looking.

a What is the theoretical probability that it is
   i red _____   **ii** yellow _____   **iii** blue? _____

Write these probabilities as decimals.

b Design an experiment to **estimate** these probabilities.
Describe what you will do and how many trials you will have.

You could put pieces of paper in a bag.

_____
_____

c Carry out your experiment, record your results on this table, and graph them.

	Frequency
Red	
Yellow	
Blue	
Total	

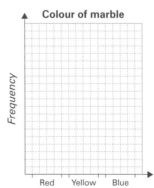

Colour of marble

Frequency

Red   Yellow   Blue

d What is your experimental probability that the marble is
   i red _____   **ii** yellow _____   **iii** blue? _____

Write these as decimals.

e Compare the theoretical and experimental probabilities. _____

_____
_____

Answers – Number

1 Working with Place Value   page 2

A
a 10^2 b 10^4 c 10^7
d 100 000   e 10 000 000   f 10 000 000 000

B
a Five hundred and ten million
b Thirty-two million, two hundred and forty-eight thousand
c Four point five times ten to the power of three
d One point five times ten to the power of eight

C
a 6252   b 1 499 997

D
a 2·44 m   b 1·533 m   c 3·164 m
d 1·88 m

2 Using Place Value to Multiply and Divide   page 3

A
a 3500   b 1 200 000   c 40
d 300   e 0·02   f 0·007
g 48   h 0·003

B
a 20   b 2000   c 200   d 2·2

C
a i 1200   ii 15 000   iii 45 000   b 0·3 kg

D
a
3·7 × 0·1 → 3·7 ÷ 100
3·7 ÷ 0·1 → 3·7 ÷ 10
3·7 × 0·01 → 3·7 × 10
3·7 ÷ 0·01 → 3·7 × 100
b
0·096 × 0·01 → 0·096 × 10
0·096 × 0·1 → 0·096 ÷ 10
0·096 ÷ 0·1 → 0·096 ÷ 100
0·096 ÷ 0·01 → 0·096 × 100

E

	6	41	0·46	3·9	0·8	⁻7·29
× 0·1	0·6	4·1	0·046	0·39	0·08	⁻0·729
÷ 0·1	60	410	4·6	39	8	⁻72·9
× 0·01	0·06	0·41	0·0046	0·039	0·008	⁻0·0729
÷ 0·01	600	4100	46	390	80	⁻729

3 Ordering Decimals and Rounding   page 4

A

START	0·69	1·34 →1·345	1·324	12·301	12·31	12·3	FINISH
	0·7 →0·71	1·453	1·354	12·03	12·311→12·32	
	0·65	0·708	1·543 →1·6 →1·61	12·13	12·09	

B
a 830   b 8400   c 8000
d 5870   e 50 000   f 300 000
g 480 000   h 2 000 000

C 16 350   16 449

D

	0 d.p.	1 d.p.	2 d.p.
4·15837	4	4·1 4·2	4·16
28·369	29 28	28·3 28·4	28·37
0·9042	0 1	0·9	0·91 0·90
169·598	171 170	169·5 169·6	169·60

E
a 53 minutes to the nearest minute
b £1897·06 to the nearest penny
or £1897 to the nearest pound

4 Adding and Subtracting Integers   page 5

A ON EVERY CONTINENT THERE IS A CITY CALLED ROME.

B
a   b

C
a 8   b 5   c ⁻7   d ⁻26
e Many answers possible.
Possible answers include
8 – 20 = ⁻12, ⁻5 – 7 = ⁻12, ⁻15 – ⁻3 = ⁻12

D One possible answer is

E
a 178   b ⁻586   c 327
d 60·4   e ⁻83   f ⁻187

5 Multiplying and Dividing Integers   page 6

A
a ⁻15, (15), ⁻15, ⁻15
b 24, 24, 24, (⁻24)
c (9), ⁻9, ⁻9, ⁻9

B
a 24 → ⁻8 → 2 → ⁻1 → 6
b ⁻6 → ⁻30 → 15 → ⁻5 → 10
c ⁻10 → 40 → 8 → ⁻4 → 1

C
a

×	3	⁻1	⁻4
⁻2	⁻6	2	8
5	15	⁻5	⁻20
⁻7	⁻21	7	28

b

×	5	⁻2	⁻9
⁻6	⁻30	12	54
⁻4	⁻20	8	36
3	15	⁻6	⁻27

c

×	3	⁻7	⁻9
⁻5	⁻15	35	45
4	12	⁻28	⁻36
⁻8	⁻24	56	72

D Many answers possible.
Possible answers include.
⁻3 × 4, ⁻12 × 1, ⁻24 ÷ 2, 36 ÷ ⁻3

E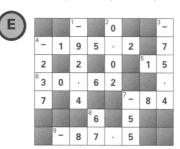

6 Order of Operations with Integers   page 7

A
- **a** 14
- **b** ⁻7
- **c** ⁻14
- **d** 401
- **e** ⁻4
- **f** 75
- **g** 108
- **h** 228
- **i** ⁻16
- **j** 42
- **k** ⁻19
- **l** ⁻2
- **m** ⁻12
- **n** ⁻3

The largest recorded iceberg was 335 km by 97 km.

B
- **a** Andy
- **b** Andy
- **c** Andy
- **d** Andy
- **e** Judy. Judy was pressing the equals key in the middle of each calculation as well as at the end, for example.

a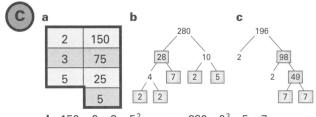

C
- **a** $4 \times {}^-3 + 2 = {}^-10$
- **b** ${}^-3 \times {}^-5 - 1 = 14$
- **c** $12 - 6 \times {}^-3 = 30$
- **d** ${}^-30 - {}^-12 \div 2 = {}^-24$
- **e** $({}^-6 + {}^-4) \times 2 = {}^-20$
- **f** $(9 - 8) \times {}^-11 = {}^-11$
- **g** ${}^-7 + 5 \times ({}^-8 - 2) = {}^-57$

7 Divisibility, Factors and Multiples   page 8

A
A 3 and 4, because 3 and 4 have no common factors other than 1.

B
- **a** 1440, 9240, 2040, 19 836
- **b** 1440, 9240, 2040, 9135
- **c** 1440, 19 836
- **d** 1440, 9240, 2040, 13 940

C

a

2	150
3	75
5	25
	5

b 280

28		10
4	7	2	5
2	2		

c 196

2	98
	2	49
	7	7

- **d** $150 = 2 \times 3 \times 5^2$
- **e** $280 = 2^3 \times 5 \times 7$
- **f** $196 = 2^2 \times 7^2$

D
a 140 ⬤ 180

HCF = 20
LCM = 1260

b 110 ⬤ 165

HCF = 55
LCM = 330

c 210 ⬤ 132

HCF = 6
LCM = 4620

8 Squares and Cubes   page 9

A
- **a** Bella
- **b** Carl
- **c** Annie
- **d** Carl
- **e** Bella
- **f** Annie
- **g** Annie

B
- **a** **i** ⁻4 and 4 **ii** ⁻8 and 8 **iii** ⁻6 and 6
- **b** **i** 7 or ⁻7 **ii** 10 or ⁻10

C
- **a** $\sqrt{9 \times 25} = 3 \times 5 = 15$
- **b** $\sqrt{4 \times 49} = 2 \times 7 = 14$

D
- **a** 8
- **b** 27
- **c** 1
- **d** 4
- **e** 1000
- **f** 5
- **g** 3
- **h** 1
- **i** 10
- **j** ⁻8
- **k** 0·008

E
- **a** 70·56
- **b** 13·93
- **c** 22·09
- **d** 10·95
- **e** 1·17
- **f** ⁻4096
- **g** 15·63
- **h** 6·14

9 Adding and Subtracting Mentally   page 10

A
- **a** 6
- **b** 7
- **c** 1·8
- **d** ⁻1·1
- **e** 6·3
- **f** 13
- **g** 1·39
- **h** 0·25
- **i** 150
- **j** 2
- **k** 422
- **l** 3711
- **m** 15·93
- **n** 2·33

B
D is the odd one out because in all of the other diagrams the outside four numbers add up to the centre number.

C

a

1	11	21	31
29	23	9	3
27	17	15	5
7	13	19	25

b

1·7	4·2	6·7	9·2
8·7	7·2	3·7	2·2
8·2	5·7	5·2	2·7
3·2	4·7	6·2	7·7

c

⁻7	⁻2	3	8
7	4	⁻3	⁻6
6	1	0	⁻5
⁻4	⁻1	2	5

D
- **a** 23 and 11
- **b** 1·8 and 2·1
- **c** 4 and ⁻7
- **d** 102 and 205

10 Multiplying and Dividing Mentally   page 11

A
THERE ARE ABOUT TWICE AS MANY BICYCLES AS MOTORCARS IN THE WORLD

B
- **a** 2·43
- **b** 16·8
- **c** 1·58
- **d** 169

C
- **a** $\boxed{5} \times \boxed{130} + \boxed{120} = 770$ or $\boxed{130} \times \boxed{5} + \boxed{120} = 770$
- **b** $\boxed{2·4} \times \boxed{3·5} + \boxed{1·6} = 10$ or $\boxed{3·5} \times \boxed{2·4} + \boxed{1·6} = 10$
- **c** $\boxed{152} \div \boxed{8} - \boxed{6} = 13$
- **d** $\boxed{0·1} \times \boxed{4·2} + \boxed{0·3} \times \boxed{0·5} = 0·57$
  (Numbers being multiplied may be reversed.)

11 Solving Problems Mentally   page 12

A
- **a** 8200 g
- **b** 1002 mm
- **c** 120 hours
- **d** 49
- **e** 26
- **f** 60°
- **g** 30p or £0·30
- **h** 15

B
- **a** $43 \times 2 = 86$ $34 \times 2 = 68$ $32 \times 4 = 128$
  $23 \times 4 = 92$ $24 \times 3 = 72$
- **b** 114
- **c** 25
- **d** £47·50
- **e** **i** 27 **ii** 2·25 **iii** 49·5
- **f** Eight 20p coins

12 Making Estimates   page 13

A
a exact number **b** estimate **c** exact number
d estimate

B
a Julia **b** Christine **c** Julia

C
Possible answers are
a $400 \times 30 = 12\,000$ **b** $800 \div 40 = 20$
c $300 \times 60 = 18\,000$
d $15 \times 4 = 60$ or $10 \times 4 = 40$
e $56 \div 7 = 8$ **f** $300 \div 6 = 50$
g $2 \times 200 = 400$ m **h** $20 \times 700 = 14\,000$
i £300 ÷ 20 = £15

**13 Adding and Subtracting – Written
Calculations   page 14**

A
a 52·01 **b** 30·13 **c** 31·29
d 8·89 **e** 32·78

B
a 32·88 kg **b** 2·65 ℓ

C
a
```
  3.2
  4.07
+ 5.96
  13.23
```
b
```
  0.49
  6.3
+ 29.19
  35.98
```
c
```
  70.16
- 16.27
  53.89
```
d $64·\boxed{5} + 3·8\boxed{2} + 1\boxed{2} = \boxed{8}0·32$

14 Multiplying – Written Calculations   page 15

A
a 253·4 **b** 2185·5 **c** 19·968

B
a

×	36	0·8	2·3
29	1044	23·2	66·7
607	21852	485·6	1396·1
27·5	990	22	63·25

b

×	0·6	3·5	8·7
2·6	1·56	9·1	22·62
71·3	42·78	249·55	620·31
6·54	3·924	22·89	56·898

C
£54·93

D
A = 2, B = 1, C = 7, D = 8

15 Dividing – Written Calculations   page 16

A
a 19·5 **b** 2·8 **c** 650 **d** 168·3

B
THE WORLD'S FIRST WOMAN PRIME MINISTER
(IN CEYLON).

16 Checking Answers   page 17

A
a Yes; because $3 \times £2 = £6$, so his change will be
   about £4.
b No, because 589 is less than 600, and if 600 is
   divided into three piles there would be 200
   sweets in each. Therefore the answer should be
   less than 200.
c No; because 51 is greater than any of the four
   numbers.

B
a false **b** true **c** true
d false **e** true **f** true

C
Answers may vary. Possible answers are
a $3·75 + 16·49 = 20·24$
b $308·2 \div 134 = 2·3$
c $\sqrt{96·04} = 9·8$
d ✓ **e** (35·82) **f** (114·8)

D
a B; because $30 \times 40 = 1200$ so the answer will be
   less than 1200. Also the last digit will be 2.
b C; because the answer must be greater than
   563, and it ends in 2.

17 Using the Calculator   page 18

A
a She did not put a right bracket after the 6.
b He did not put brackets around 8×6.

B

1	2 3	·	3 4		4 1
	5		5 7	2	5
6 8	·	7	2		·
	4			·	6
7 2	9	·	9		3
			8 6	8	

C
a 90p or £0·90 change
   Possible keying sequence is

   0 STO M+ 4 × 12·37
   STO M+ 3 × 16·54
   M+ 100 − RCL M+ =

b 5·5 km
   Possible keying sequence is

   0 STO M+ 3 × 5·7
   STO M+ 4 × 4·9
   M+ 42·2 − RCL M+ =

18 Writing Fractions   page 19

A
a i $\frac{3}{8}$ **ii** $\frac{5}{9}$ **iii** $\frac{7}{12}$
b i Red: about $\frac{1}{4}$ Silver: about $\frac{1}{8}$
   ii Petroleum: about $\frac{2}{5}$ Other: about $\frac{1}{8}$

B
a $\frac{3}{4}$ **b** $\frac{1}{2}$ **c** $\frac{3}{20}$
d $\frac{1}{8}$ **e** $\frac{1}{9}$ **f**
g $\frac{2}{3}$ **h** $\frac{7}{8}$

C
a $\frac{1}{6}$ **b** $\frac{17}{24}$ **c** $\frac{1}{5}$ **d** $\frac{4}{15}$ **e** $\frac{2}{5}$

19 Fractions and Decimals   page 20

A
1 a $\frac{2}{5}$ **b** $\frac{7}{20}$ **c** $\frac{367}{1000}$ **d** $\frac{2}{25}$
   e $\frac{17}{20}$ **f** $\frac{3}{5}$ **g** $\frac{61}{125}$ **h** $\frac{8}{125}$
   i $6\frac{1}{5}$ **j** $3\frac{71}{100}$
2 a 0·8 **b** 0·12 **c** 0·45 **d** 0·625
   e 0·3 **f** 0·58 **g** 2·375 **h** 1·2
3 a 0·22 **b** 6·33 **c** 0·43
4 a 0·22 **b** 0·82 **c** 1·24 **d** 0·60

B
$0·18 \quad \frac{18}{200}$

C

Terminating	Recurring
$\frac{3}{12}$ $\frac{18}{45}$ $\frac{21}{16}$	$\frac{2}{3}$ $\frac{16}{22}$ $\frac{60}{81}$ $\frac{17}{132}$

 D Many answers are possible. One possible answer is:

$\frac{5}{8}$ $\frac{7}{14}$ $\frac{9}{15}$ $\frac{11}{55}$ $\frac{21}{24}$

20 Ordering Fractions Page 21

 A **a** $\frac{1}{6} = \frac{1}{6}$ $\boxed{\frac{1}{3} = \frac{2}{6}}$ **b** $\frac{1}{4} = \frac{2}{8}$ $\boxed{\frac{3}{8} = \frac{3}{8}}$

c $\boxed{\frac{4}{5} = \frac{8}{10}}$ $\frac{7}{10} = \frac{7}{10}$ **d** $\frac{2}{3} = \frac{8}{12}$ $\boxed{\frac{3}{4} = \frac{9}{12}}$

e $\boxed{\frac{2}{5} = \frac{16}{40}}$ $\frac{3}{8} = \frac{15}{40}$ **f** $\boxed{\frac{13}{15} = \frac{26}{30}}$ $\frac{5}{6} = \frac{25}{30}$

g $\frac{5}{9} = \boxed{0.5\overline{5}}$ $\boxed{\frac{10}{17} = 0.59}$

h $\boxed{\frac{2}{11} = 0.1\overline{8}}$ $\frac{3}{19} = \boxed{0.16}$

i $\frac{6}{7} = \boxed{0.86}$ $\boxed{\frac{19}{21} = 0.90}$

B **a** CDs **b** France **c** Sophie

C **a**

[number line from 0 to 2 with arrows at $\frac{1}{5}$, $\frac{6}{15}$, $\frac{40}{25}$ above and $\frac{7}{20}$, $\frac{36}{30}$ below]

b

[number line from 0 to $\frac{50}{100}$ with arrows at $\frac{1}{20}$, $\frac{8}{40}$, $\frac{3}{10}$, $\frac{9}{20}$, $\frac{6}{15}$]

c **i** $\frac{7}{16}, \frac{1}{2}, \frac{5}{8}, \frac{3}{4}$ **ii** $\frac{3}{10}, \frac{7}{20}, \frac{2}{5}, \frac{1}{2}, \frac{3}{5}$

 D **a** $\frac{13}{24}$ **b** $\frac{13}{16}$

21 Adding and Subtracting Fractions Page 22

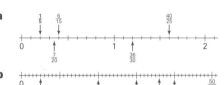 **A** **a** $\boxed{\frac{1}{2} + \frac{1}{3}} \rightarrow \boxed{\frac{5}{6} - \frac{1}{6}} \rightarrow \boxed{\frac{2}{3} - \frac{1}{12}} \rightarrow \boxed{\frac{7}{12} + \frac{5}{12}} \rightarrow \boxed{1}$

b $\boxed{\frac{3}{4} + \frac{1}{8}} \rightarrow \boxed{\frac{7}{8} - \frac{1}{2}} \rightarrow \boxed{\frac{3}{8} - \frac{1}{8}} \rightarrow \boxed{\frac{1}{4} + \frac{2}{5}} \rightarrow \boxed{\frac{13}{20} - \frac{3}{10}} \rightarrow \boxed{\frac{7}{20}}$

c $\boxed{\frac{2}{3} + \frac{1}{5}} \rightarrow \boxed{\frac{13}{15} - \frac{1}{3}} \rightarrow \boxed{\frac{8}{15} - \frac{2}{5}} \rightarrow \boxed{\frac{2}{15} + \frac{7}{30}} \rightarrow \boxed{\frac{11}{30} - \frac{3}{10}} \rightarrow \boxed{\frac{1}{15} + \frac{2}{3}} \rightarrow \boxed{\frac{11}{15}}$

B **a** $\frac{5}{12}$ **b** $\frac{4}{15}$

C **a**

$\frac{2}{8}$	$\frac{5}{8}$	$\frac{5}{8}$
$\frac{7}{8}$	$\frac{4}{8}$ or $\frac{1}{2}$	$\frac{1}{8}$
$\frac{3}{8}$	$\frac{3}{8}$	$\frac{6}{8}$

b

$\frac{2}{5}$	$\frac{7}{10}$	$\frac{7}{10}$
$\frac{9}{10}$	$\frac{3}{5}$	$\frac{3}{10}$
$\frac{1}{2}$	$\frac{1}{2}$	$\frac{4}{5}$

c

$\frac{1}{6}$	$\frac{7}{12}$	$\frac{1}{2}$
$\frac{3}{4}$	$\frac{5}{12}$	$\frac{1}{12}$
$\frac{1}{3}$	$\frac{1}{4}$	$\frac{2}{3}$

 D Answers will vary. Possible answers include
$\frac{1}{2} = \frac{2}{5} + \frac{1}{10}$ $\frac{1}{2} = \frac{6}{7} - \frac{5}{14}$ $\frac{1}{2} = \frac{1}{4} + \frac{1}{6} + \frac{1}{12}$

22 Multiplying and Dividing Integers by Fractions Page 23

A $\frac{1}{2} \times 32 = 16, \frac{1}{2} \times 34 = 17, \frac{1}{2} \times 36 = 18, \frac{1}{2} \times 38 = 19,$
$\frac{1}{3} \times 30 = 10, \frac{1}{3} \times 33 = 11, \frac{1}{3} \times 36 = 12, \frac{1}{3} \times 39 = 13$

B **a** 20 **b** 24 **c** 6 **d** 14

C THE ELECTRIC CHAIR WAS INVENTED BY A DENTIST.

23 Fractions, Decimals and Percentages Page 24

A

Fraction	Decimal	Percentage
$\frac{4}{5}$	0·8	80%
$\frac{3}{5}$	0·6	60%
$\frac{2}{5}$	0·4	40%
$\frac{9}{20}$	0·45	45%
$\frac{3}{100}$	0·03	3%

Fraction	Decimal	Percentage
$\frac{3}{25}$	0·12	12%
$\frac{21}{30}$	0·7	70%
$1\frac{18}{25}$	1·72	172%
$2\frac{13}{20}$	2·65	265%
$\frac{1}{3}$	0·$\overline{3}$	$33\frac{1}{3}$%

B **a** 42% **b** 63% **c** 154%
d 58·5% **e** 46·2% **f** 45·7%
g 66·1% **h** 675·4%

C **a** 20% **b** $66\frac{2}{3}$% **c** 40%
d 95% **e** $87\frac{1}{2}$%

D **a** 5% **b** Schist, 22%
c Gneiss, $12\frac{1}{2}$% **d** Slate, 40%

24 Finding Percentages Page 25

A **a** 14 **b** 11 **c** 72
d 152 **e** 123 **f** 121
g 104 **h** 94 **i** 27·9
A cricket ball has between 120 and 150 stitches.

B **a** 14·4 cm **b** £57·60
c 26·66 kg **d** £94·50

C **a** £9440 **b** 120·9 m^2 **c** 33·28 m^2

25 Percentage Increase and Decrease Page 26

A **a** £25·60 **b** £76·50
c £25·20 **d** £7·22

B **a** Wrong; because if they increased by 100% we would add 100% on. £35 + £35 = £70.
b Wrong; because if they increased by 300% we would add 300% on. 300% = £315. The necklace is now worth £105 + £315 = £420.
c C or 1·3

C **a** Thursday **b** £21·15

D **a** £138 **b** £151·80 **c** £86·25

26 Direct Proportion   Page 27

Eliza's Shopping List		Carl's Shopping List	
8 roses	£24	5 roses	£15
3 camelias	£13·50	7 camelias	£31·50
14 petunias	£9·10	3 petunias	£1·95
7 lavenders	£19·25	2 lavenders	£5·50
5 gnomes	£62·50	9 gnomes	£112·50
12 statues	£550·20	14 statues	£641·90
Total cost	£678·55	**Total cost**	£808·35

a 24 g   **b** 28 g   **c** 0·4 ℓ or 400 mℓ

20

27 Simplifying Ratios   Page 28

a The ratio is not written in its simplest form. 3 : 10.
b Ratios should not have decimals in them. 5 : 4.
c The ratio is written using different units (£ and p). 19 : 8

a 1 : 3   **b** 2 : 5   **c** 2 : 3 : 4
d 7 : 12 : 5   **e** 7 : 4   **f** 9 : 8
g 21 : 40   **h** 3 : 10   **i** 3 : 5
j 13 : 200   **k** 1 : 30   **l** 50 : 23

C
a 9 : 5   **b** 19 : 33

D
a 4 cm and 6 cm   **b** 14 mm and 20 mm

28 Ratio and Proportion   Page 29

A
a **i** 4 : 3   **ii** 6 : 5 : 8
b $\frac{8}{19}$ **c** $\frac{5}{19}$ **d** $\frac{3}{4}$

B
a **i** 8 : 5   **ii** 3 : 4   **iii** 5 : 3 : 8
b **i** 0·4   **ii** 0·15   **iii** 0·2

C
a 1 : 2   **b** 12 : 11 : 18   **c** $\frac{11}{50}$
d 18%   **e** 60%

29 Solving Ratio and Proportion Problems   Page 30

A
a 720 mℓ   **b** 120 g   **c** 8 cups   **d** 12 t

B
a 20 m   **b** 120 m   **c** 242 m   **d** 384 m

C
240 cats

D
a 30 m   **b** 37·5 m

E
a 30 people   **b** 6 years old

30 Dividing in a Given Ratio   Page 31

A
a **i** £24   **ii** £300
    iii £496   **iv** £79·40
b **i** £50   **ii** £140
    iii £335   **iv** £4077·50

B
a Nitrogen 200 g, Phosphate 250 g, Potassium 450 g
b 2 ℓ

C
a 30°, 150°   **b** 30°, 45°, 105°
c 40°, 60°, 120°, 140°

D
Overall winner was King's College by 7 games.

Answers – Algebra

31 Getting Started with Algebra   Page 33

A

Expressions
$3x - 1$
$\frac{6m - 3}{2}$
$4a + c$
$\frac{7e - d}{8}$

Equations
$c + d = 8$
$5x - 1 = 7$
$6e + f = 12$
$\frac{2p - q}{3} = 7$

Formulae
$W = 6D$
$S = 5t^2$
$V = l^3$
$D = \frac{m}{v}$

Functions
$y = 2x + 1$
$y = \frac{4x - 1}{3}$
$y = \frac{x}{8} + 6$
$y = \frac{x + 4}{5}$

B
a Expression; because there is no equals sign; yes
b Formula; because it has more than one unknown and each letter stands for something specific

C
a $2a$ **b** ab **c** a
d $3(a + b)$ **e** $2a^2$ **f** $\frac{2a + b}{2}$
g $5ab$ **h** $3a + b$ **i** $2ab$

32 Understanding Algebra   Page 34

A
a $14 - \boxed{3x} = 5$ **b** $5(\boxed{x - 1}) = 10$
c $\frac{15(\boxed{x - 2})}{3} = 5$ **d** $8 + 4\boxed{x^2} = 44$

B
a abc means $a \times b \times c$. $a \times b \times c$ does not equal $a + b + c$.
b $105 - 28$ does **not** equal $105 - 30$
$105 - 28 = 105 - 30 + 2$
$= 75 + 2$
$= 77$

C
a i T   ii T   iii F   iv F   v T
b Answers will vary. Possible answers are:

i $b = c + d$
$b = 3d$
ii $d = b - c$
$d = \frac{c}{2}$
iii $2a = 4c$
$2a = 2b + c$
iv $3c = a + 2d$
$3c = 2b$

D
a $8 - d = c,\ d = 8 - c$ **b** $\frac{q}{3} = p,\ 3 = \frac{q}{p}$
c $4 = \frac{m}{n},\ 4n = m$ **d** $\frac{y + 1}{4} = x$
e $a = 3b - 2$

33 Simplifying Expressions   Page 35

A THE HEART OF A BLUE WHALE IS THE SIZE OF A SMALL CAR.

B
a
b
c

C

a	b	c	d
$2 \times 2m$	$c \times c \times c$	$a^2 \times 6$	$h \times 5h$
$\frac{8m}{2}$	$c^9 \div c^3$	$\frac{12a^2}{2}$	$\frac{15h^2}{3}$
$4 \times m^2$	$c^5 \div c^2$	$\frac{18a}{3}$	$\frac{20h^2}{4h}$
$\frac{24m}{6}$	$\frac{c^6}{c^3}$	$3a \times 2a$	$\frac{10h^3}{2h}$

34 Brackets   Page 36

A
a $5c + 10$ **b** $3b - 24$ **c** $2a - 24$
d $6p + 18$ **e** $4s - 20$ **f** $3a + 3$
g $2y - 12$ **h** $^-2d - 16$ **i** $^-4c + 16$
j $ab - ac$ **k** $15g + 5h$ **l** $33y - 110z$

B
a ✓   **b** ✗   **c** ✗   **d** ✗
e ✓   **f** ✓   **g** ✗   **h** ✗
i ✓   **j** ✓   **k** ✗   **l** ✓

C

$2(n - 5)$	$2n$	$7n$	$9n$	$3(6n - 2)$	$18n$	$^-8$	$24n$
$^-10$	$7n$	$7(n - 2)$	$^-10$	^-15n	$^-6$	$4(6n - 2)$	15
7	$7(n + 1)$	$^-14$	^-9n	$^-5(3n + 2)$	$^-50$	$21n$	$3(7n + 5)$
$12n$	$4(3n + 2)$	8	$^-9(n - 3)$	$^-27$	$25(3n - 2)$	$75n$	75
$2(6n - 4)$	$12n$	$^-8$	27	$3(2n - 9)$	$6n$	$15(3n + 5)$	$45n$

The expression left is $9n$.

35 Collecting Like Terms   Page 37

A
a $4n + 5x - 2x + 3n = 7n + \boxed{3x}$
b $8a - 4a + 7b - 5b = \boxed{4a} + \boxed{2b}$
c $9x - 3a - 6x + 5a = \boxed{3x} + \boxed{2a}$
d $12b - 5c - 7b - 2c = \boxed{5b - 7c}$
e $4c + 4 + 2c - 2 + c = \boxed{7c + 2}$
f $9y + 2 - 3y + 6 - y = \boxed{5y + 8}$
g $5x^2 + 2x^2 + x^2 = \boxed{8x^2}$
h $3x^2 + 2x + 4x^2 = \boxed{7x^2 + 2x}$
i $6n + 2x + \boxed{2n} + \boxed{3x} = 8n + 5x$
j $5c + 4d - \boxed{2c} - \boxed{10d} = 3c - 6d$
k $6a - \boxed{5a} + \boxed{3} + 8 = a + 11$
l $3x^2 + \boxed{3x^2} + \boxed{5x} + 2x = 6x^2 + 7x$
m $2(3a - 2) + 4(2 - a) = \boxed{2a + 4}$
n $12 + 3(x + 2) = \boxed{18 + 3x}$
o $4(n + 3) - 2(n + 5) = \boxed{2n + 2}$
p $6(3n + a) - (4n + 2a) = \boxed{14n + 4a}$

B Answers may vary. Possible answers are
a $P = 2x + 3 + 3y + 2x + 3 + 3y$
$= 4x + 6y + 6$
b $P = 10x + 7 + y - 4 + 10x + 7 + y - 4$
$= 20x + 2y + 6$
or
$P = 2(10x + 7) + 2(y - 4)$
$= 20x + 2y + 6$
c $P = 5(x + 2) + 4(x - 1) + 5(x + 2) + 4(x - 1)$
$= 18x + 12$
or
$P = 10(x + 2) + 8(x - 1)$
$= 18x + 12$

C **a**

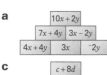

10x + 2y

7x + 4y	3x − 2y

4x + 4y	3x	⁻2y

b

11a + 5b

7a + 3b	4a + 2b

3a + 3b	4a	2b

c

c + 8d

3c + 6d	2d − 2c

4d	3c + 2d	⁻5c

d

5q − 2p

2q	3q − 2p

2q − 6p	6p	3q − 8p

C **a** Yes; because 11y is the area of the large rectangle and 8 is the area of the missing piece.
b $9y + 2(y − 4)$
c 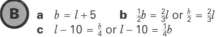 $11(y − 4) + 9 × 4 = 11(y − 4) + 36$
d Simplify **b**: $9y + 2(y − 4) = 9y + 2y − 8 = 11y − 8$
Simplify **c**: $11(y − 4) + 36 = 11y − 44 + 36 = 11y − 8$

36 Substituting into Expressions   Page 38

A **a** 9   **b** 7
c 6   **d** 6
e 4   **f** 42
g 45   **h** 54
i 3   **j** 76
k 20   **l** 74
m 30   **n** ⁻1
o 12   **p** 3
q 2   **r** 8
s 14   **t** 4·5
u 21   **v** 36
w 1·4   **x** 2
y ⁻2   **z** 4

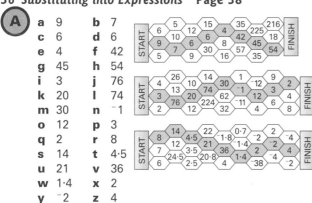

B **a** 35   **b** 143   **c** 399
d 2499

C

2	6	7
10	5	0
3	4	8

Yes; square A is magic, because every row, column and diagonal adds to 15.

37 Substituting into Formulae   Page 39

A **a i** 2   **ii** 2·5   **iii** 4·75
b i 180 mℓ **ii** 270 mℓ **iii** 369 mℓ
c i A = 15 cm^2 **ii** A = 20·74 cm^2
d i £27·50   **ii** £68·80

B **a i** $b = 12$ cm   **ii** $b = 7$ mm
b i 3   **ii** 14   **iii** 29   **iv** 36
c i 4   **ii** 7   **iii** 26
d i $l = 2$ m   **ii** $l = 6$ m   **iii** 13 m

38 Writing Expressions   Page 40

A **1 a** $15x$ **b** $5x − 3$
2 a $3p$ **b** $4p$ **c** $4p − 3$
3 a $\frac{t}{2}$ or $\frac{1}{2}t$ **b** $\frac{t}{2} + 45$ or $\frac{1}{2}t + 45$
4 a $3y$ **b** $3y + 4$ **c** $2(3y + 4)$ or $6y + 8$
5 a $2n + 400$ **b** $3n + 320$
c $n + \frac{2}{3}n + 200$ or $1\frac{2}{3}n + 200$
d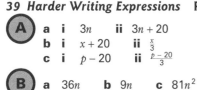

39 Harder Writing Expressions   Page 41

A **a i** $3n$ **ii** $3n + 20$
b i $x + 20$ **ii** $\frac{x}{3}$
c i $p − 20$ **ii** $\frac{p − 20}{3}$

B **a** $36n$ **b** $9n$ **c** $81n^2$

40 Writing Equations   Page 42

A **1 a** $f + v = 48$ **b** $f + 4 = 16$ **c** $v = 3f$
2 a $h + t = £80$ **b** $10t = £450$ **c** $h + 10 = t$

B **a** $b = l + 5$ **b** $\frac{1}{2}b = \frac{2}{3}l$ or $\frac{b}{2} = \frac{2}{3}l$
c $l − 10 = \frac{b}{4}$ or $l − 10 = \frac{1}{4}b$

C **1 a** $t = 4p$ **b** $t + 500 = 6p$
c $\frac{p}{5} = \frac{t}{20}$ or $\frac{1}{5}p = \frac{1}{20}t$
2 a $f + a = p$ **b** $a − 4 = \frac{f}{5}$ **c** $\frac{f}{3} + a = \frac{p}{2}$

41 Writing and Finding Formulae   Page 43

A **a** $m = 1000n$ **b** $s = 2c$
c $t = 4 − h$ **d** $w = 400b + 900$

B **a**
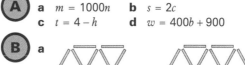
4 pens   5 pens
b

Number of pens, p	1	2	3	4	5	...
Number of fence sections, f	3	5	7	9	11	...

c $f = 2p + 1$

C **a**
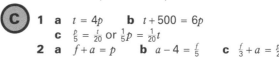
b

Number of purple tiles, p	1	2	3	4	5	...
Number of white tiles, w	6	8	10	12	14	...

c $w = 2p + 4$

42 Writing and Solving Equations   Page 44

A **a** $x = 4$ **b** $y = 6$ **c** $z = 7$ **d** $n = 6$
e $a = 19$ **f** $d = 3$ **g** $w = 11$ **h** $y = 8$
i $c = 3·2$ **j** $n = 2$ **k** $p = 5$ **l** $m = 3$
m $p = 9$ **n** $x = 8$ **o** $t = 2·4$

B **a** Cost of 1 muffin $m − 1$
Multiply by 8 $8 × (m − 1)$
Equation $8(m − 1) = 16$
$8m − 8 = 16$
$8m = 16 + 8$
$8m = 24$
$m = \frac{24}{8}$
$m = 3$
Muffins cost £3 each.
b $x + x + 9 + 7x = 45$
$9x + 9 = 45$
$9x = 45 − 9$
$9x = 36$
$x = \frac{36}{9}$
$x = 4$
Robert has 4 CDs.
Stephen has 13 CDs.
Terry has 28 CDs.

C **a**

$2n + 30 = 60$
$n = 15$

b

$2n + 38 = 84$
$n = 23$

c

$2n + 47 = 85$
$n = 19$

43 Solving Equations by Transforming Both Sides page 45

A **a** $x = 12$ **b** $w = 4$ **c** $n = 2$

B ON AVERAGE HICCUPS LAST FIVE MINUTES.

C **a** $2n + 5 = 3n - 30$ **b** $5n + 2 = 3n + 46$
$n = 35$ $n = 22$
c $4n - 7 = 23 - n$ **d** $2n + 4n - 6 = 2(2n - 1) + n$
$n = 6$ or $6n - 6 = 5n - 2$
$n = 4$

44 Equations and Graphs page 46

A **a**

Number of hours worked	1	2	3	4	5	...
Money earned (£)	5	10	15	20	25	...

b The ratio is 1 : 5 for all pairs of values.
c Yes; because the ratio is constant.
d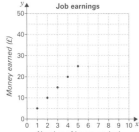

Yes; the points lie on a straight line.
e $y = 5x$
f £45. Using the formula $y = 5x$ and substituting 9 for x or extending the line of points on the graph to (9, 45).

B **a**

Yellow paint	5	10	15	20	25	...
Red paint	3	6	9	12	15	...

b Yes; 5 : 3.
c

d C $y = \frac{3}{5}x$
e **i** 21 spoons **ii** 90 spoons

45 Generating Sequences page 47

A 4, 7, 10, 13, 16, 19, 22

B **a** 1, 3, 9, 27, 81
b 625, 125, 25, 5, 1
c 4, ⁻8, 16, ⁻32, 64
d 16, 4, 1, $\frac{1}{4}$, $\frac{1}{16}$ or 16, 4, 1, 0·25, 0·0625
e 80, 160 **f** 6, 3 **g** 81, ⁻243

C **a** 4, 5, 7, 10, 14, 19
b 20, 22, 26, 32, 40, 50
c 50, 49, 46, 41, 34, 25

D **a** 1, 2, 4, 8, 16, 32
b **i** 2, 3, 5, 8, 12, 17 **ii** 7th day

46 Continuing Sequences page 48

A

a 2, 7, 12, 17, ... $a = 4, d = 2$
b 4, 6, 8, 10, ... $a = ⁻2, d = 4$
c ⁻2, 2, 6, 10, 14, ... $a = 2, d = 5$
d 4, 2, 0, ⁻2, ⁻4, ... $a = 4, d = ⁻2$
e 5, 9, 13, 17, ... $a = 5, d = 6$
f 5, ⁻1, ⁻7, ⁻13, ... $a = 5, d = ⁻6$
g 5, 1, ⁻3, ⁻7, ... $a = 5, d = 4$
h 5, 11, 17, 23, ... $a = 5, d = ⁻4$

B **a** 50, 55, 60, 65, 70, 75
b 0·3, 0·4, 0·5, 0·6, 0·7, 0·8
c ⁻10, ⁻7, ⁻4, ⁻1, 2, 5
d 3, 2, 1, 0, ⁻1, ⁻2

C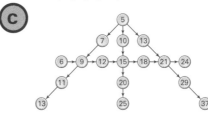

D **a** ⁻3 **b** 1, 2, 4, 5, 10 or 20

E **a** 35, 42, 49 equal
b 48, 96, 192 unequal
c 186, 182, 178 equal
d 18, 24, 31 unequal
e 8, 4, 2 unequal
f 26, 42, 68 unequal
g Two possible answers are
● 2, 3, 5, 8, 12, 17 add 1, 2, 3, 4, 5, ...
● 2, 3, 5, 8, 13, 21 add the two previous terms

47 Writing Sequences from Rules page 49

A **a** 4, 10, 16, 22, 28, 34, 40
b 5, 50, 500, 5000, 50 000, 500 000
c 96, 48, 24, 12, 6, 3, 1·5
d 7, 8, 11, 16, 23, 32
e 1, 4, 5, 9, 14, 23, 37
f 6, 8, 12, 18, 26, 36

B

$T(n)$	$T(1)$	$T(2)$	$T(3)$	$T(4)$	$T(5)$	...	$T(20)$
$2n + 1$	3	5	7	9	11		41
$15 - n$	14	13	12	11	10		⁻5
$35 - 3n$	32	29	26	23	20		⁻25
$\frac{1}{2}n + 3$	$3\frac{1}{2}$	4	$4\frac{1}{2}$	5	$5\frac{1}{2}$		13

$T(n) = 3n + 1$ • — • first term 2, rule add 1
<u>2, 3, 4, 5, 6</u>

$T(n) = 1 + n$ • — • first term 1, rule add 2
<u>1, 3, 5, 7, 9</u>

$T(n) = 2n - 1$ • — • first term 4, rule add 3
<u>4, 7, 10, 13, 16</u>

$T(n) = 20 - 2n$ • — • first term 1·5, rule add 2
<u>1·5, 3·5, 5·5, 7·5, 9·5</u>

$T(n) = 2n - 0·5$ • — • first term 18, rule subtract 2
<u>18, 16, 14, 12, 10</u>

D a Two possible answers are
- First term 1, rule add 4
- First term 3, rule add 4

b First term 44, rule subtract 10

48 Describing linear sequences   page 50

A $T(n) = 3n$ • — • multiples of 3, starting at ⁻3
$T(n) = 3n + 1$ • — • multiples of 3, starting at 6
$T(n) = 3n + 3$ • — • descending numbers starting at 4, with a difference of 3 between terms
$T(n) = 3n - 6$ • — • multiples of 3, starting at 3
$T(n) = 7 - 3n$ • — • ascending numbers starting at 4, with a difference of 3 between consecutive terms, all one more than a multiple of 3

B a

Difference between terms is odd
$T(n) = 5n - 3$ $T(n) = 7n + 4$
$T(n) = n + 12$ $T(n) = 8 - 3n$ $T(n) = 10 - n$

Difference between terms is even
$T(n) = 2n + 4$ $T(n) = 6n - 1$
$T(n) = 7 - 2n$

b

Terms are multiples of 6
$T(n) = 6n$ $T(n) = 6n + 6$
$T(n) = 18 - 6n$

Terms are not multiples of 6
$T(n) = 6n + 3$

C a Ascending sequence starting at 3 with a difference of 4 between terms, all one less than a multiple of 4.

b Descending sequence starting at 60 with a difference of 6 between terms, all multiples of 6.

D a Many possible answers.
Possible answers include $T(n) = 7n + \boxed{7}$
$T(n) = 7n + \boxed{42}$
$T(n) = 7n + \boxed{⁻14}$
The number in the box must be a multiple of 7.

b $T(n) = 2n$ c $T(n) = 2n + 3$ d $T(n) = 110 - 10n$

49 Sequences in Practical Situations   page 51

A a   b 3, 6, 9, 12, ...   Add 3

c Multiply the carton number, n, by 3; $3n$

B a

Cake size	1	2	3	4	5
Number of liquorice strips	4	8	12	16	20
Number of sweets	5	9	13	17	21
Total number of decorations	9	17	25	33	41

b Multiply n by 4; $4n$
c 4 are added each time plus the first cake has 1 extra; $4 \times n + 1$ or $4n + 1$.
d $8n + 1$

C a

Shape number	1	2	3
Number of stars on outside	6	24	54

b There are n^2 stars on each face, and 6 faces on each cube. So you multiply n^2 by 6.
c $6n^2$

50 Finding the rule for the nth term   page 52

A a Bob b Cathy c Ann

B a

Term number	1	2	3	4	5
$T(n)$	4	8	12	16	20
Difference		4	4	4	4

nth term = **$4n$**

b

Term number	1	2	3	4	5
$T(n)$	3	7	11	15	19
Difference		4	4	4	4

nth term = **$4n - 1$**

c

Term number	1	2	3	4	5
$T(n)$	2	8	14	20	26
Difference		6	6	6	6

nth term = **$6n - 4$**

d

Term number	1	2	3	4	5
$T(n)$	0·2	0·4	0·6	0·8	1·0
Difference		0·2	0·2	0·2	0·2

nth term = **$0·2n$**

e

Term number	1	2	3	4	5
$T(n)$	50	45	40	35	30
Difference		⁻5	⁻5	⁻5	⁻5

nth term = **$55 - 5n$**

f

Term number	1	2	3	4	5
$T(n)$	2	⁻4	⁻10	⁻16	⁻22
Difference		⁻6	⁻6	⁻6	⁻6

nth term = **$8 - 6n$**

C a 601 b 1020 c 802
d The difference between terms is 8. From this we get the rule for the nth term as $T(n) = 8n + 2$. We then substitute in $n = 100$ to get $T(100) = 8 \times 100 + 2 = 802$.

D Yes; because the rule for the sequence is $9n - 2$. If $9n - 2 = 205$ then $n = 23$, so the 23rd door is numbered 205.

51 Functions   page 53

 a

Input	Output
2	9
5	21
10	41
12	49

b

Input	Output
33	3
63	6
83	8
18	1·5

 a
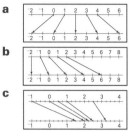

b

c

> Lines on a mapping diagram will be parallel
> $x \rightarrow x+3$ $x \rightarrow x-3$

> Lines on a mapping diagram will meet on
> the zero line when extended backwards
> $x \rightarrow 3x$ $x \rightarrow 5x$ $x \rightarrow \frac{1}{3}x$

52 More Functions   page 54

 a
$2, 4, 3, 5, 1 \rightarrow$ [add 5] $\rightarrow 7, 9, 8, 10, 6$

b
$1, 3, 2, 5, 4 \rightarrow$ [multiply by 3] $\rightarrow$ [add 4] $\rightarrow 7, 13, 10, 19, 16$

c
$6, 4, 8, 2, 10 \rightarrow$ [multiply by 2] $\rightarrow$ [subtract 3] $\rightarrow 9, 5, 13, 1, 17$

d
$9, 5, 1, 7, 3 \rightarrow$ [multiply by 3] $\rightarrow$ [subtract 1] $\rightarrow 26, 14, 2, 20, 8$

 a
$4, 9, 3, 6 \rightarrow$ [multiply by 2] $\rightarrow$ [add 2] $\rightarrow 10, 20, 8, 14$

$4, 9, 3, 6 \rightarrow$ [add 1] $\rightarrow$ [multiply by 2] $\rightarrow 10, 20, 8, 14$

b

$16, 24, 8, 40 \rightarrow$ [divide by 4] $\rightarrow$ [add 3] $\rightarrow$

$\rightarrow$ [subtract 3] $\rightarrow$ [multiply by 4] $\rightarrow 16, 24, 8, 40$

 a Equivalent   **b** Not equivalent

 a
$\underline{7}, \underline{4} \leftarrow$ [divide by 3] $\leftarrow$ [add 4] $\leftarrow 17, 8$
The inverse function of
$x \rightarrow 3x - 4$ is $x \rightarrow \frac{x+4}{3}$

b
$\underline{18}, \underline{33} \leftarrow$ [subtract 2] $\leftarrow$ [multiply by 5] $\leftarrow 4, 7$
The inverse function of
$x \rightarrow \frac{x+2}{5}$ is $x \rightarrow \mathbf{5x - 2}$

53 Graphing Linear Functions   page 55

a (1, 4), (2, 8)   **b** (2, 3), (6, 7)
c (2, 7), (5, 13)   **d** (3, 8), (⁻2, ⁻7)
e (5, 7), (4, 8)   **f** (1, ⁻5), (2, ⁻12)

 a (2, 7), (1, 5), (0, 3), (⁻1, 1)   **b**
c **i** No   **ii** Yes   **iii** Yes
d No; because $2 \times \mathbf{15} + 3$ does not equal 35.
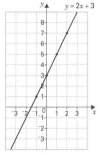

 a

x	0	1	2	3	⁻1
y	4	2	0	⁻2	6

b
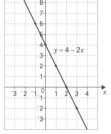

c Yes; because $4 - 2 \times \mathbf{8} = 4 - 16 = \mathbf{⁻12}$, so the point (8, ⁻12) is on the line $y = 4 - 2x$.
d (4, ⁻4), (⁻2, 8), (1·5, 1), (⁻0·5, 5), (3·5, ⁻3)

54 Graphing Linear Sequences   page 56

 a

Shape number	1	2	3	4
Number of toothpicks	4	7	10	13

b (1, 4), (2, 7), (3, 10), (4, 13)
c
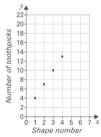

d Yes
e No; because you cannot have a term number that is not a whole number.
f 22
g No; because 20 is not part of the sequence 4, 7, 10, 13.

 a

Term number	1	2	3	4
$T(n)$	4	2	0	⁻2

b
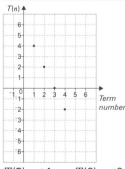

c $T(5) = ⁻4$ $T(6) = ⁻6$

d No; because it wouldn't lie in a straight line with the other points on the graph **OR** because $6 - 2n$ cannot have a value of $^-5$ if n is a whole number.

e $T(n) = 6 - 2n$ is a set of points on an imagined straight line and $y = 6 - 2x$ is a straight line.

55 Equations of Straight-line Graphs   page 57

A
a $y = 6x$ **b** $y = 3 - 5x$ and $y = 2 - 3x$
c $y = x + 3$ and $y = 3 - 5x$ and $y = 3$
d $y = x - 2$ **e** $y = 3$

B

Equation	$y = 4x + 3$	$y = 2x - 1$	$y = \frac{1}{2}x + 2$	$y = {}^-5x$	$y = 10 - 3x$	$* y = 3x - 1$	$* y = \frac{1}{4}x$
Gradient	4	2	$\frac{1}{2}$	$^-5$	$^-3$	3	$\frac{1}{4}$
Coordinates of y-intercept	(0, 3)	(0, $^-1$)	(0, 2)	(0, 0)	(0, 10)	(0, $^-1$)	(0, 0)

C
a B   **b** D   **c** G   **d** E   **e** F   **f** C

D
Many possible answers.
Possible answers include
a $y = \frac{1}{3}x$, $y = \frac{1}{3}x + 4$
b $y = x - 2$, $y = 2x - 2$
c $y = 4x + 1$, $y = 20x + 1$

56 Reading Real-life Graphs   page 58

A
a Glasgow
b

Location	Average high temperature	
	°F	°C
London, UK	**70°**	21°
Bangkok, Thailand	95°	**35°**
Kathmands, Nepal	86°	**30°**
Death Valley, USA.	**116°**	47°

Answers are approximate.
c About 57°C   **d** About $^-14$°C

B
a About 2·8 kg   **b** About 11·2 kg
c

Becky's length	Becky's weight
58 cm	6 kg
66·5 cm	8·4 kg
53·5 cm	5 kg

d About 1·2 kg

57 Plotting Real-life Graphs   page 59

A
a

Length of fabric (m)	1	2	3	4	8
Cost (£)	£2·50	£5	£7·50	£10	£20

b (3, 7·50), (4, 10), (8, 20)
c
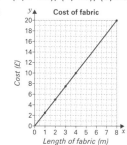
Cost of fabric

d **i** About £15   **ii** About £8·75   **iii** About 7 m

B
a JORDAN

Number of months	0	1	2	3	4
Jordan's savings (£)	20	60	100	140	180

MICHAEL

Number of months	0	1	2	3	4
Michael's savings (£)	50	80	110	140	170

b
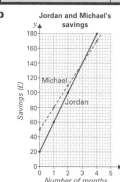
Jordan and Michael's savings

c After 3 months.
d Jordan

58 Distance/Time Graphs   page 60

A
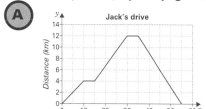
Jack's drive

B

Olivia's adventure

C
a 260 km   **b** 4 hours 15 mins
c 45 mins   **d** 160 km

59 Interpreting Real-life Graphs   page 61

A D

B
a As distance from the city centre increases the price of a house decreases.
b The bar is lifted steadily, then held, then lifted again, then held and dropped.
c As time since dawn increases, the temperature stays the same, then rises, then falls, then rises more than before, then falls again.

C
a Ben   **b** Bradley; 170 cm
c 12 years and 17 years   **d** Bradley

60 Sketching and Interpreting Real-life Graphs
page 62

  a B   **b** D   **c** A   **d** C

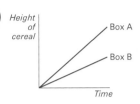

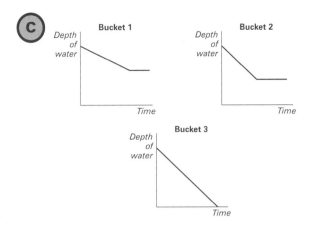

Answers – Shape, Space and Measures

61 Angles Made with Intersecting and Parallel Lines
page 64

Complementary angles	Supplementary angles	Neither
A C H	B E I	D F G

a $d = 120°$; because corresponding angles on parallel lines are equal.
b $d = 75°$; because alternate angles on parallel lines are equal.

		Diagram 1	Diagram 2
a	2 pairs of corresponding angles	h and d, g and c, f and b, a and e	e and g, d and f
b	2 pairs of alternate angles	a and g, d and f	d and h, a and e c and i, b and l
c	2 angles that are equal to angle a	c, e, g	g, e
d	4 pairs of supplementary angles	f and e, g and h, e and h, f and g, a and d, b and c, ...	d and e, b and c, i and l, j and k, a and f, h and g, ...

Many possible answers for **d**.

D **a** PQ and RS are parallel because the two 79° angles are alternate and alternate angles on **parallel lines** are equal.
b DE and IG are parallel because the two 63° angles are corresponding and corresponding angles on **parallel lines** are equal.

62 Using Geometric Reasoning to Find Angles
page 65

a
shaded angle + 30° + 50° = 180°
 (Angles in a △ add to 180°.)
shaded angle = 180° − 30° − 50°
 = 100°
p = shaded angle = 100°
 (Corresponding angles, parallel lines are equal.)
b 66° + 90° + p = 180°
 (Angles on a straight line add to 180°.)
p = 180° − 66° − 90°
 = 24°

B **a** • $a + 45° + 85° = 180°$
 (Angles on a straight line add to 180°.)
$a = 180° − 45° − 85°$
 = 50°
• $b = a = 50°$
 (Corresponding angles on parallel lines are equal.)
• $c = 45°$ (Alternate angles on parallel lines are equal.)
b • $d + 130° = 180°$ (Angles on a straight line add to 180°.)
$d = 180° − 130°$
$d = 50°$

• shaded angle + 110° = 180°
 (Angles on a straight line add to 180°.)
shaded angle = 180° − 110°
shaded angle = 70°
e = shaded angle
 (Corresponding angles on parallel lines are equal.)
$e = 70°$
• $d + e + f = 180°$ (Angles in a triangle add to 180°)
$50° + 70° + f = 180°$
$f = 180° − 50° − 70°$
$f = 60°$

C **a** $x + 15° = 81°$ (Vertically opposite angles are equal)
$x = 81° − 15°$
$x = 66°$
b $2x + 15° = 4x − 17°$
 (Corresponding angles on parallel lines are equal.)
$15 + 17 = 4x − 2x$
$32 = 2x$
$x = 16°$

63 Angles in Triangles   page 66

A **a** 118° **b** 126° **c** 52° **d** 58° **e** 65°

B **a** $x = 53°$ (Exterior angles of a triangle are equal to the sum of the two opposite interior angles.)
$y = 75°$ (Angles on a straight line add to 180°.)
b $x = 70°$ (Base angles in an isosceles triangle are equal and angles in a triangle add to 180°.)
$y = 110°$ (Exterior angles of a triangle are equal to the sum of the two opposite interior angles.)

C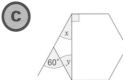
$x = 30°$
In a regular hexagon x and y are equal and $x + y = 60°$, so $x = 30°$

D **a** $a = 76°$ **b** $c = 18°$ **c** $e = 34°$
$b = 138°$ $d = 24°$ $f = 34°$

64 Angles in Quadrilaterals   page 67

A **a** 110° **b** 80° **c** 143° **d** 138°
e 198° **f** 82° **g** 68°

B
a $c + 118° + 21° = 180°$ (Angles in a △ add to 180°)
$c = 41°$
$p + 41° = 180°$
 (Angles on a straight line adds to 180°)
$p = 180° − 41°$
 (Corresponding angles are equal)
 = 139°
$q = 41°$

b

- $d + 48° + 90° = 180°$ (Angles in a △ add to 180°.)
  $d = 42°$
- $r = 42$ (Vertically opposite angles.)
- $e + 72° = 180°$ (Angles on a straight line.)
  $e = 108°$
- $s + 37° + 108° + 42° = 360°$
  (Angles in a quadrilateral add to 360°.)
  $s = 360° - 37° - 108° - 42°$
  $s = 173°$

C

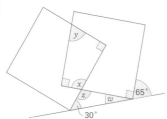

We name the unknown angles w and z.

- $w + 90° + 65° = 180°$
  (Angles on a straight line add to 180°.)
  $w = 180° - 90° - 65°$
  $w = 25°$
- $z + 25° + 30° = 180°$ (Angles in a △ add to 180°.)
  $z = 180° - 25° - 30°$
  $z = 125°$
- $x = z = 125°$ (Vertically opposite angles are equal.)
- $y + 90° + 90° + 125° = 360°$
  (Angles in a quadrilateral add to 360°.)
  $y = 360° - 90° - 90° - 125°$
  $y = 55°$

65 Visualising and Sketching 2-D Shapes   page 68

A   **a**   Possible answers are
   i

squares   rectangles   isosceles   right–angled
                                triangles   triangles

   ii   Possible answers are

kite   parallelogram   trapezium

   b   Possible answers are

       or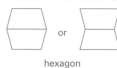

hexagon   parallelogram

c   Possible answers are

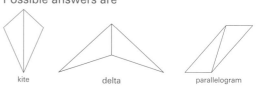

kite   delta   parallelogram

B   Answers will vary.
The shape is a parallelogram which has its shorter diagonal drawn in.
The parallelogram is divided into two scalene triangles by the diagonal.
An isosceles triangle is attached to the parallelogram. One of the equal sides of the isosceles triangle is formed by the shorter side of the parallelogram.

66 Properties of Triangles and Quadrilaterals   page 69

A   **a**   Isosceles triangle   **b**   Rhombus
   c   Kite or arrowhead

B   **a**   True   **b**   False   **c**   True   **d**   False
   e   True; because all parallelograms have two pairs of parallel sides and all rhombuses have this.
   f   False; because all squares have four 90° angles and not all rhombuses do; only **some** rhombuses are squares.

C   **a**   ∠X = 60° because all angles of an equilateral triangle are equal.
      XY = 4 cm because all sides of an equilateral triangle are equal.
   b   ∠X = 98° because opposite angles of a parallelogram are equal.
      XY = 30 mm because opposite sides of a parallelogram are equal.

D   **a**   Two pairs of parallel sides   **OR**   diagonals bisect each other
   b   One line of symmetry

67 Tessellations   page 70

A   **b**   Yes   **c**   No

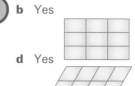 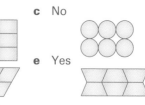

   d   Yes   **e**   Yes

 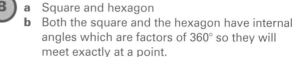

   f   No

B   **a**   Square and hexagon
   b   Both the square and the hexagon have internal angles which are factors of 360° so they will meet exactly at a point.

C a Answers will vary. Two possible answers are

 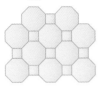

b

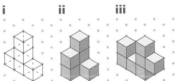

68 Congruence   page 71

A a

D and M	F and L
A and Y and V	I and O
B and X	J and T
C and N	P and H
E and K and U	Q and S

b G, R, W

B a No; because they are not the same size.
   b No; because the corresponding angles are not equal.

C b **c** **d**

scalene triangle   right angled triangles   isosceles triangles

D a DEF and VWX
   b ∠F and ∠X
   c EF and XV

69 Describing and Sketching 3-D Shapes and Nets   page 72

A a 9   **b** 3   **c** 6

B a i   ii   iii
   b

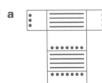 

C a

b

c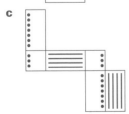

D E and K, X and C, L and P

70 Plans and Elevations   page 73

A a
front	F	**b** front	B
side	A	side	I
plan	H	plan	C

c
front	G	**d** front	B
side	B	side	G
plan	D	plan	E

B

front   plan   side

C

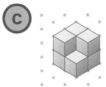

D A possible answer is

 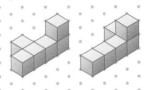

Same side view

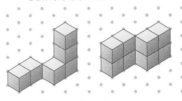

Same plan view

71 Construction   page 74

A a

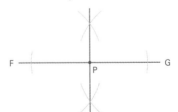

b

B a

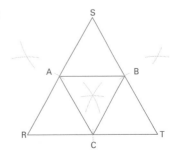

b 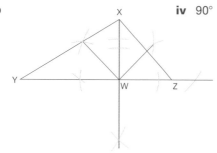 **iv** 90°

72 Constructing Triangles and Quadrilaterals
page 75

(A) **a** About 5·7 cm   **b** About 54°

(B) **a** About 117°   **b** About 129°

(C) **a** About 8·8 cm   **b** About 91°

73 Loci page 76

A **a** A circle

b

A line half way between GH and IJ, and parallel to them.

c   A line which is the perpendicular bisector of the line joining T1 and T2.

d A line which is the bisector of ∠QRS.

e

74 Coordinates and Transformations page 77

A **1 a** A = (1, 5) B = (4, 5) C = (4, 3) D = (1, 1)

b i A′ = (1, ⁻5) B′ = (4, ⁻5) C′ = (4, ⁻3)
    D′ = (1, ⁻1)

ii A′ = (⁻1, 5) B′ = (⁻4, 5) C′ = (⁻4, 3)
    D′ = (⁻1, 1)

iii A′ = (⁻1, ⁻5) B′ = (⁻4, ⁻5) C′ = (⁻4, ⁻3)
    D′ = (⁻1, ⁻1)

2 a   **b** (4, ⁻3)

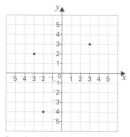

c i (4, 6), (⁻2, 5), (⁻1, ⁻1), (5, 0)
ii (⁻3, 3), (⁻2, ⁻3), (3, 4), (4, ⁻2)
iii (⁻3, 3), (3, 2), (2, ⁻4), (⁻4, ⁻3)

Shape, Space and Measures

3 a (1, 4), (⁻3, ⁻2), (2, ⁻3)
b i (0, 6), (⁻4, 0), (1, ⁻1)
ii (4, ⁻1), (⁻2, 3), (⁻3, ⁻2)
iii (1, ⁻6), (⁻3, 0), (2, 1)

75 Combinations of Transformations page 78

(A) **1 a**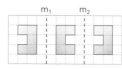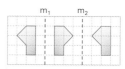

b translation

2 a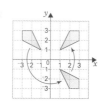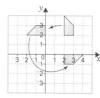

b a half-turn rotation

(B) **a**
b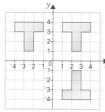

c 1 unit right and 3 units down

(C) **a** 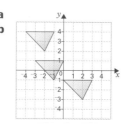 **c**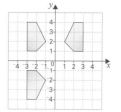

b 270°; the origin   **d** 270°; (1, 1)

76 Symmetry page 79

(A)

	a	b	c	d	e
Number of lines of symmetry	2	1	0	1	0
Order of rotation symmetry	2	1	4	1	6

(B) **a** If a regular hexagon is rotated, the angles fit on top of each other six times as it is rotated.
b As a rectangle is rotated the two parts of the diagonals fit onto each other twice so they must bisect each other.

(C)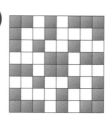

D Possible answers include

a   b

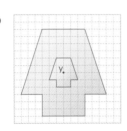

c   or

77 Enlargement   page 80

A   **a** 2   **b** 3   **c** 2

B   **a**   **b**

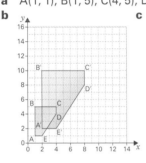

c When a shape is enlarged the **angles** stay the same but the **lengths** change.

C   **a**   A(1, 1), B(1, 5), C(4, 5), D(4, 4), E(2, 1)

b   **c**

A'(2, 2), B'(2, 10), C'(8, 10), D'(8, 8), E'(4, 2)

A'(3, 3), B'(3, 15), C'(12, 15), D'(12, 12), E'(6, 3),

d The coordinates of ABCDE are each multiplied by the scale factor, for example B(1, 5) × scale factor 3 = B'(3, 15).

78 Scale Drawing   page 81

A   **a** About 0·9 m or 90 cm   **b** About 3·6 m
    c About 2·7 m   **d** About 9 m

B   **a** 4·5 m by 4·5 m   **b** 3 m by 6 m
    c 1·5 m by 7·5 m

C   **a**   **i** 900 m   **ii** 1800 m or 1·8 km
        iii 3600 m or 3·6 km
    b   **i** 2500 cm   **ii** 25 m
    c 1 cm represents 10 m.
    d 1 cm represents 500 m, or 1 cm represents $\frac{1}{2}$ km.

D

79 Finding the Mid-point of a Line   page 82

A   **a** (4, 3)   **b** (5, 5)   **c** (⁻1, 6)   **d** (1·5, 8)

B   **a** (4, 4)   **b** (6, ⁻1·5)

C   **a**   **i** (3, 6)   **ii** (⁻3·5, 3)
    b (1, ⁻4·5)

80 Metric Conversions   page 83

A   IN A FEW MILLION YEARS THERE WON'T BE A LEAP YEAR.

B   **a**   **i**   24 years, 4 months
        ii   19 hours, 13 minutes
    b   **i**   4 Feb 2004   **ii**   11 March 2004

C   **a** 36 ha   **b** 37 500 cm³
    c   **i** 98 800 kg   **ii** 98·8 tonnes
    d 32 sacks.

81 Metric and Imperial Equivalents   page 84

A   **a** Dominic   **b** Dominic   **c** Bella
    d Bella   **e** Carl   **f** Dominic
    g Carl   **h** Dominic

B   Possible answers are:
    a   **i** 10 cm   **ii** 50 cm   **iii** 2 m   **iv** 5 m
    b   **i** 8·8 lb   **ii** 6·6 lb   **iii** 5·5 lb   **iv** 9·9 lb
    c   **i** 9 ℓ   **ii** 1·2 ℓ or 1200 mℓ
        iii 12 ℓ   **iv** 11·25 ℓ

C   **a** First: John, Second: Nathan, Third: Antony, Fourth: Kishan
    b Deptford, Stockbridge, Ringwood, Upavon

82 Units, Measuring Instruments and Accuracy   page 85

A   **a** C   **b** L   **c** E   **d** H
    e M   **f** A   **g** C or D   **h** I
    i J   **j** B   **k** F or G   **l** N or O

B   **a** A   **b** C   **c** F

C   **a** A   **b** C   **c** D or E

83 Estimating   page 86

A   Answers will vary. Possible answers are
    a About 10–15 mℓ (2 or 3 teaspoons).
    b About 2·4–3 m (a bit higher than a door).
    c About 600–800 g (less than a large bag of sugar).
    d About 300–450 mℓ (more than 1 glass and less than 2 glasses).
    e About 3–5 tonnes (3–5 small cars).
    f About 120–180 cm² (quarter of a piece of A4 paper)

B Answers will vary. Possible answers are
a 80 cm–1 m
b 40–70 cm
c 200–400 mℓ
d 200–500 g
e 50–150 g
f 15–25 ℓ
g 50 secs–1$\frac{1}{2}$ minutes
h 8–12 mins
i 3000–4000 cm^2 or 0·3–0·4 m^2

84 Bearings   page 87

A a East   b South   c North east
d 270°   e 135°

B a i 071°   ii 251°   b i 114°   ii 294°
c i 262°   ii 082°   d i 317°   ii 137°

C a 53°   b 158°   c 234°   d 338°

D a This diagram is not full size.

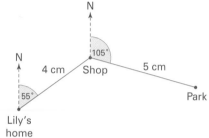

b About 83°

85 Area and Perimeter   page 88

A a A = 54 m^2 P = 36 m   b 120 cm^2
c A = 24·5 m^2 P = 23·9 m   d 20 mm^2
e 12·5 cm^2

B a A = 36 mm^2 b A = 9 cm^2
c A = 48 m^2, P = 32 m   d h = 7 cm
e A = 30 m^2 f A = 11·84 m^2
g A = 5·145 m^2 P = 12·2 m

C a 49 cm^2
b Possible answers include:
40 mm and 12 mm
10 mm and 48 mm
 5 mm and 96 mm
16 mm and 30 mm
64 mm and 7·5 mm

86 Volume and Surface Area   page 89

A a i   V = 8 m^3 SA = 24 m^2
ii   V = 12 000 cm^3 SA = 3800 cm^2
iii   V = 0·312 m^3 SA = 3·34 m^2
b i 12   ii 72   iii 3072 cm^2

C L 480 cm^3 E 570 cm^3 H 903 cm^3

D a 396 m^2 b 20 m, 6 m, 3 m

E Pupils' own work

Answers – Handling Data

87 Discrete and Continuous Data   page 91

A
a C   b D   c C   d C   e D
f C   g C   h D   i D   j C

B a

Time in supermarket (mins)	Tally	Frequency
$0 \leqslant t < 10$	IIII	5
$10 \leqslant t < 20$	II	2
$20 \leqslant t < 30$	I	1
$30 \leqslant t < 40$	IIII	4
$40 \leqslant t < 50$	IIII	4
$50 \leqslant t < 60$	II	2

b 4   c 7   d 10

C a

Distance travelled (km)	Tally	Frequency
$0 \leqslant t < 50$	III	3
$50 \leqslant t < 100$	IIII	4
$100 \leqslant t < 150$	IIII	4
$150 \leqslant t < 200$	IIII	5
$200 \leqslant t$	II	2

b 4
c 16
d These are five class intervals on the table. The intervals $0 \leqslant t < 10$, $10 \leqslant t < 20$, ... give 20 intervals for this data. This is too many to be able to see the distribution of distances clearly.

88 Two-way Tables   page 92

A a 12   b 14   c 34

B a 4   b 12   c Cricket   d 41

C a 4   b 14   c 15   d 47

D

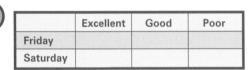

	Excellent	Good	Poor
Friday			
Saturday			

89 Surveys – Collecting the Data   page 93

Answers will vary. Possible ideas are
1   a • How many hours per week do children play sport?
    • What sports do children play?
    • Is it different for different ages, or for girls and boys?
  b • Most children play sport for at least 1 hour per week.
    • Girls play more sport than boys.
    • Football is the most popular sport.
  c Sports played, hours spent playing, age, gender.
  d • Male ☐   Female ☐
    • Age: 5–7 ☐   8–10 ☐   11–13 ☐
    • Sport played:
      Football ☐   Rugby ☐   Hockey ☐
      Netball ☐   Cricket ☐   Tennis ☐
      Other ☐

• Hours spent playing sport (including practice):
    0 up to 3 ☐   3 up to 6 ☐
    6 up to 9 ☐   9 or more ☐
  e at least 50
  f Primary

2   Possible answers are:
  a • Do attractions with cheaper entrance costs attract more visitors?
    • Are attractions in large cities more popular?
  b • Historical sites, like castles, are popular.
    • Higher numbers of tourists visit less expensive attractions.
    • Historical tourist attractions in London are more popular than modern tourist attractions.
  c Number of visitors to attraction, location, cost, type of attraction.
  d

Attraction name	Number of visitors per year	Entrance cost	Location	Type: Historical or modern

  e About 50 historical and 50 modern.
  f Secondary

90 Mode and Range   page 94

A a 1   b 4

B a i 1600 $\leqslant d <$ 1800   ii 1800 $\leqslant d <$ 2000
  b 991 m   c 1099·7
  d So she could see whether the pupil's fitness had improved

C a i 87·6 g   ii 24 g
  b Variety A has a much greater range showing that the mass varies more. Variety B is more consistent in mass.

91 Mean   page 95

A 1·97

B a 13   b 7·5   c 8·48

C a 23   b 24·4   c 23·7; yes

92 Median, Mean, Mode, Range   page 000

A a 166·5 mm   b 166·5
  c 169·58 mm   d 60 mm

B a i 8:00 a.m. Mean: 14·1 mph
            Median: 14 mph   Range: 13 mph
    ii 8:00 p.m. Mean: 30·3 mph
            Median: 30 mph   Range: 19 mph
  b Evening, because there is less traffic.

C **a** 2   **b** 10 or 0   **c** 4   **d** 4 or 9

93 Finding the Median, Range and Mode from a Stem-and-leaf Diagram   page 97

A **a** 28 years   **b** 25 years   **c** 21 years

B **a** 27 min   **b** 48 min   **c** 27 min

C **a**

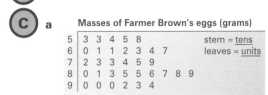

Masses of Farmer Brown's eggs (grams)

5	3 3 4 5 8
6	0 1 1 2 3 4 7
7	2 3 3 4 5 9
8	0 1 3 5 5 6 7 8 9
9	0 0 0 2 3 4

stem = tens
leaves = units

b 9   **c** 12   **d** 75 g   **e** 41 g   **f** 90 g

94 Comparing Data   page 98

A Possible answers are:
- The M road because the mean and the median are slightly shorter, and the range is large so sometimes the trip will be very short.
- The A road because it is only slightly longer on average, and the times are much more consistent so the trip time is more predictable.

B **a** Neptune's: Mean = 630 g   Range = 120 g
Cousteau's: Mean = 640 g   Range = 41 g
b Possible answer is:
Cousteau's; because the mean mass of chips is higher than at Neptune's, and the range is lower which means that the mass of chips is more consistent.

C **a**

	Mean	Median	Mode	Range
Nellie	2·2 m	2·2 m	2·2 m	0·1 m
Leah	2·1 m	2·3 m	2·35 m	1·3 m

b Possible answer is:
Leah; because her median and modal jump length are longer. Her mean jump length is only lower than Nellie's because of her 1·1 m jump, and this could have been a mistake.

95 Compound Bar Charts and Line Graphs   page 99

A **a** and **b**

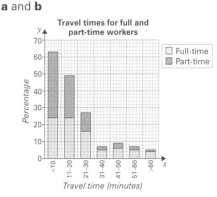

Travel times for full and part-time workers

☐ Full-time
☐ Part-time

c Full-time

B **a**

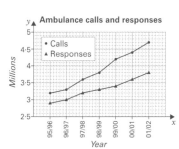

Ambulance calls and responses

• Calls
▲ Responses

b Both the number of calls made and the number of responses have increased from 1995 to 2002. The calls made have increased more, which indicates a lower proportion of calls are being responded to.

96 Frequency Diagrams   page 100

A **a**

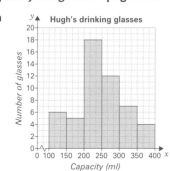

Hugh's drinking glasses

b 4   **c** 11   **d** 52

B **a**

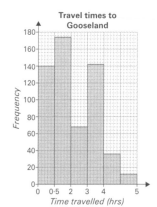

Travel times to Gooseland

b 314
c 48
d No; because the second interval on the graph groups travel times between $\frac{1}{2}$ an hour and one hour. We cannot tell how many of these people travelled for less than $\frac{3}{4}$ of an hour.
e 0·5–
f Possible answers are:
- Travel times might affect opening times, or timings of shows.
OR • She might want to know how far away she needs to advertise Gooseland.

97 Drawing Pie-Charts   page 101

A B and D

B **a** **ii** 108°   **iii** $\frac{4}{30} \times 360 = 48°$ **iv** $\frac{5}{30} \times 360 = 60°$

b

Fruit gum flavours

Lime

Orange

Strawberry

Lemon

C **a**

Music style	Number of CDs	Pie chart angle
Jazz	8	64°
Rock	17	136°
Classical	5	40°
Hip Hop	11	88°
Country	4	32°
Total	45	360°

b

Kieran's CD collection

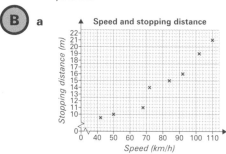

Country

Jazz

Hip hop

Rock

Classical

98 Scatter Graphs   page 102

A **a**

Time watching TV and doing homework

Time doing homework (hours) vs *Time watching TV (hours)*

b Incorrect; the graph shows that the more time you spend watching T.V., the less homework you do.

B **a**

Speed and stopping distance

Stopping distance (m) vs *Speed (km/h)*

b The stopping distance increases.

99 Interpreting graphs   page 103

A **a** Muesli bar
b The muesli bar has a greater proportion of carbohydrate, and a lower proportion of fat than the crisps.

B **a** **i** about 60%   **ii** about 25%   **iii** about 15%
b KS2 pupils get more help from family than KS3 pupils.
KS3 pupils get more help from friends than KS2 pupils.

C **a** In 1993 about 18% of men did shift work. This increased to just over 25% in 1998, then slowly dropped to about 18% again in 2003.
b Men: About 16%   Women: About 17%

101 Language of Probability   page 105

A **a** C   **b** **i** B   **ii** D   **iii** C

B B, because grid B has a greater proportion of even numbers than grids A and C.

C **a** ■   **b** ●   **c** ◆

D Bag 1, because there is a smaller proportion of banana sweets in bag 1 than in bag 2.

102 Calculating Probability   page 106

A **a** 5%   **b** 0·6   **c** $\frac{1}{7}$

B **a**

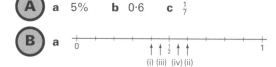

0 ↑ ↑ $\frac{1}{2}$ ↑ ↑ 1
(i) (iii) (iv) (ii)

b **i** Because there are not the same number of red hats and blue hats.
ii 8

C **a** $\frac{1}{2}$ **b** $\frac{1}{2}$

103 Calculating Probability by Listing Outcomes   page 107

A **a** black/red, black/yellow, black/green, black/blue, white/red, white/yellow, white/green, white/blue
b

Pete	swim	swim	swim	shop	shop	shop	tennis	tennis	tennis
Jenny	swim	shop	tennis	swim	shop	tennis	swim	shop	tennis

B **a**

Baby 1	girl	girl	boy	boy
Baby 2	girl	boy	girl	boy

b **i** $\frac{1}{4}$ **ii** $\frac{1}{4}$ **iii** $\frac{1}{2}$

C **a**

Spinner

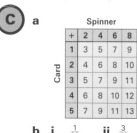

+	2	4	6	8
1	3	5	7	9
2	4	6	8	10
3	5	7	9	11
4	6	8	10	12
5	7	9	11	13

Card

b **i** $\frac{1}{20}$ **ii** $\frac{3}{20}$ **iii** $\frac{1}{10}$
iv $\frac{11}{20}$ **v** $\frac{11}{20}$ **vi** $\frac{2}{5}$

104 Estimating Probability from Experiments
page 108

 a $\frac{44}{47}$ **b** $\frac{35}{47}$

B **a** $\frac{1}{4}$ **b** $\frac{35}{100}$ or $\frac{7}{20}$ **c** $\frac{90}{300}$ or $\frac{3}{10}$

C **i** $\frac{1}{5}$ **ii** $\frac{16}{25}$

D **a** Katie; because $\frac{1}{10}$ is a much higher proportion than $\frac{2}{100}$.

    b Alice's; because she repeated her experiment 100 times, compared to Katie who only did 10 trials.

105 Comparing Calculated Probability with
Experimental Probability **page 109**

 a $\frac{1}{2}$

    d no

    e closer to $\frac{1}{2}$ heads than in the experiment with only 50 throws.

B **a** **i** 0·5   **ii** 0·3   **iii** 0·2